Surface Engineering of Light Alloys

Surface Engineering of Light Alloys

Editor

Sara Ferraris

MDPI • Basel • Beijing • Wuhan • Barcelona • Belgrade • Manchester • Tokyo • Cluj • Tianjin

Editor
Sara Ferraris
Politecnico di Torino
Italy

Editorial Office
MDPI
St. Alban-Anlage 66
4052 Basel, Switzerland

This is a reprint of articles from the Special Issue published online in the open access journal *Coatings* (ISSN 2079-6412) (available at: https://www.mdpi.com/journal/coatings/special_issues/surf_eng_light_alloy).

For citation purposes, cite each article independently as indicated on the article page online and as indicated below:

LastName, A.A.; LastName, B.B.; LastName, C.C. Article Title. *Journal Name* **Year**, *Volume Number*, Page Range.

ISBN 978-3-0365-0120-8 (Hbk)
ISBN 978-3-0365-0121-5 (PDF)

Cover image courtesy of Sara Ferraris.

© 2021 by the authors. Articles in this book are Open Access and distributed under the Creative Commons Attribution (CC BY) license, which allows users to download, copy and build upon published articles, as long as the author and publisher are properly credited, which ensures maximum dissemination and a wider impact of our publications.

The book as a whole is distributed by MDPI under the terms and conditions of the Creative Commons license CC BY-NC-ND.

Contents

About the Editor vii

Sara Ferraris
Special Issue: Surface Engineering of Light Alloys
Reprinted from: *Coatings* **2020**, *10*, 1177, doi:10.3390/coatings10121177 1

Fabio Alemanno, Veronica Peretti, Angela Tortora and Silvia Spriano
Tribological Behaviour of Ti or Ti Alloy vs. Zirconia in Presence of Artificial Saliva
Reprinted from: *Coatings* **2020**, *10*, 851, doi:10.3390/coatings10090851 3

Mateusz Niedźwiedź, Władysław Skoneczny and Marek Bara
Influence of Conditions for Production and Thermo-Chemical Treatment of Al_2O_3 Coatings on Wettability and Energy State of Their Surface
Reprinted from: *Coatings* **2020**, *10*, 681, doi:10.3390/coatings10070681 13

Mateusz Niedźwiedź, Władysław Skoneczny and Marek Bara
The Influence of Anodic Alumina Coating Nanostructure Produced on EN AW-5251 Alloy on Type of Tribological Wear Process
Reprinted from: *Coatings* **2020**, *10*, 105, doi:10.3390/coatings10020105 29

Arezoo Ghanbari, Fernando Warchomicka, Christof Sommitsch and Ali Zamanian
Investigation of the Oxidation Mechanism of Dopamine Functionalization in an AZ31 Magnesium Alloy for Biomedical Applications
Reprinted from: *Coatings* **2019**, *9*, 584, doi:10.3390/coatings9090584 39

Seiji Yamaguchi, Phuc Thi Minh Le, Morihiro Ito, Seine A. Shintani and Hiroaki Takadama
Tri-Functional Calcium-Deficient Calcium Titanate Coating on Titanium Metal by Chemical and Heat Treatment
Reprinted from: *Coatings* **2019**, *9*, 561, doi:10.3390/coatings9090561 53

Hao Wan, Shuai Zhao, Qi Jin, Tingyi Yang and Naichao Si
The Formation of Microcrystal in Helium Ion Irradiated Aluminum Alloy
Reprinted from: *Coatings* **2019**, *9*, 516, doi:10.3390/coatings9080516 69

Martin Buchtík, Michaela Krystýnová, Jiří Másilko and Jaromír Wasserbauer
The Effect of Heat Treatment on Properties of Ni–P Coatings Deposited on a AZ91 Magnesium Alloy
Reprinted from: *Coatings* **2019**, *9*, 461, doi:10.3390/coatings9070461 79

Sara Ferraris, Sergio Perero and Graziano Ubertalli
Surface Activation and Characterization of Aluminum Alloys for Brazing Optimization
Reprinted from: *Coatings* **2019**, *9*, 459, doi:10.3390/coatings9070459 89

Martina Cazzola, Sara Ferraris, Enrico Prenesti, Valentina Casalegno and Silvia Spriano
Grafting of Gallic Acid onto a Bioactive Ti6Al4V Alloy: A Physico-Chemical Characterization
Reprinted from: *Coatings* **2019**, *9*, 302, doi:10.3390/coatings9050302 103

Jinshu Xie, Jinghuai Zhang, Shujuan Liu, Zehua Li, Li Zhang, Ruizhi Wu, Legan Hou and Milin Zhang
Hydrothermal Synthesis of Protective Coating on Mg Alloy for Degradable Implant Applications
Reprinted from: *Coatings* **2019**, *9*, 160, doi:10.3390/coatings9030160 121

About the Editor

Sara Ferraris received her PhD in Biomedical Engineering in 2010 from Politecnico di Torino. She is Assistant Professor in the Department of Applied Science and Technology at Politecnico di Torino. Her main research interests deal with surface properties of metallic materials for different applications (biomedical, mechanical, industrial), surface modification strategies (e.g., functionalization, coating, chemical or physical treatments), and surface characterizations. She is the author of 102 papers in international journals and holds 5 patents in the materials science field.

Editorial

Special Issue: Surface Engineering of Light Alloys

Sara Ferraris

Department of Applied Science and Technology, Politecnico di Torino, C.so Duca degli Abruzzi 24, 10129 Torino, Italy; sara.ferraris@polito.it

Received: 27 November 2020; Accepted: 30 November 2020; Published: 1 December 2020

Light alloys (mainly aluminum, magnesium and titanium alloys) are of great interest in applications where lightweight has an high impact, such as automotive, aerospace and biomedical fields.

In addition to their bulk properties, such as mechanical properties, thermal properties or density, surface properties are crucial in many applications. In fact the surface is the first layer of the material exposed to the working environment. In biological systems this means that the surface is the first approach of the material to biological entities and it plays a crucial role in situations like tissue integration or bacterial contamination. Moreover the surface is a critical area for corrosion or wear phenomena both in biomedical and automotive or aerospace applications. Finally the surface can be the starting point for joining processes, in this scenario proper surface preparation can significantly affect the joint performance.

In this context surface engineering of light alloys is a versatile instrument to optimize surface properties of these materials for specific applications, without altering their bulk properties.

The present special issue covers all the above described topics with 10 research papers.

Five of them are related to biomedical applications of titanium an magnesium alloys. Surface engineering of titanium alloys moves from the investigation of Ti/Ti-alloys wear behavior in presence of artificial saliva (paper from Alemanno et al. [1]), to bioactive inorganic coatings for the improvement of bone bonding ability of Ti alloys (Yamaguchi et al. paper [2]) and finally to surface functionalization with natural molecules (Cazzola et al. paper [3]) to improve biological properties of Ti alloys. The main criticism of Mg alloys in biomedical applications is related to their poor corrosion resistance and too rapid degradation. Two solutions are explored in the present special issue, organic dopamine functionalization (Ghanbari et al. paper [4]) or inorganic coatings (Mg oxides and carbonates, Xie et al. paper [5]).

The protection of Mg alloys from corrosion and wear is of interest also for industrial applications far from the biomedical ones, Ni-P coatings are investigated for this purpose in the Buchtik paper [6].

Protective coatings (alumina) for industrial applications on Al alloys have been investigated from the wettability and mechanical standpoints in the papers from Niedźwiedź et al. in the present special issue [7,8].

The possibility to activate Al/Al-alloys surface to optimize their joining ability has been analyzed in the paper from Ferraris et al. [9].

Finally the effect of high energy irradiation on the microstructure of Al alloys is the key topic of Wan et al. paper [10].

Conflicts of Interest: The author declares no conflict of interest.

References

1. Alemanno, F.; Peretti, V.; Tortora, A.; Spriano, S. Tribological Behaviour of Ti or Ti Alloy vs. Zirconia in Presence of Artificial Saliva. *Coatings* **2020**, *10*, 851. [CrossRef]
2. Yamaguchi, S.; Thi Minh Le, P.; Ito, M.; Shintani, S.A.; Takadama, H. Tri-Functional Calcium-Deficient Calcium Titanate Coating on Titanium Metal by Chemical and Heat Treatment. *Coatings* **2019**, *9*, 561. [CrossRef]

3. Cazzola, M.; Ferraris, S.; Prenesti, E.; Casalegno, V.; Spriano, S. Grafting of Gallic Acid onto a Bioactive Ti6Al4V Alloy: A Physico-Chemical Characterization. *Coatings* **2019**, *9*, 302. [CrossRef]
4. Ghanbari, A.; Warchomicka, F.; Sommitsch, C.; Zamanian, A. Investigation of the Oxidation Mechanism of Dopamine Functionalization in an AZ31 Magnesium Alloy for Biomedical Applications. *Coatings* **2019**, *9*, 584. [CrossRef]
5. Xie, J.; Zhang, J.; Liu, S.; Li, Z.; Zhang, L.; Wu, R.; Hou, L.; Zhang, M. Hydrothermal Synthesis of Protective Coating on Mg Alloy for Degradable Implant Applications. *Coatings* **2019**, *9*, 160. [CrossRef]
6. Buchtík, M.; Krystýnová, M.; Másilko, J.; Wasserbauer, J. The Effect of Heat Treatment on Properties of Ni–P Coatings Deposited on a AZ91 Magnesium Alloy. *Coatings* **2019**, *9*, 461. [CrossRef]
7. Niedźwiedź, M.; Skoneczny, W.; Bara, M. Influence of Conditions for Production and Thermo-Chemical Treatment of Al_2O_3 Coatings on Wettability and Energy State of Their Surface. *Coatings* **2020**, *10*, 681. [CrossRef]
8. Niedźwiedź, M.; Skoneczny, W.; Bara, M. The Influence of Anodic Alumina Coating Nanostructure Produced on EN AW-5251 Alloy on Type of Tribological Wear Process. *Coatings* **2020**, *10*, 105. [CrossRef]
9. Ferraris, S.; Perero, S.; Ubertalli, G. Surface Activation and Characterization of Aluminum Alloys for Brazing Optimization. *Coatings* **2019**, *9*, 459. [CrossRef]
10. Wan, H.; Zhao, S.; Jin, Q.; Yang, T.; Si, N. The Formation of Microcrystal in Helium Ion Irradiated Aluminum Alloy. *Coatings* **2019**, *9*, 516. [CrossRef]

Publisher's Note: MDPI stays neutral with regard to jurisdictional claims in published maps and institutional affiliations.

© 2020 by the author. Licensee MDPI, Basel, Switzerland. This article is an open access article distributed under the terms and conditions of the Creative Commons Attribution (CC BY) license (http://creativecommons.org/licenses/by/4.0/).

Article

Tribological Behaviour of Ti or Ti Alloy vs. Zirconia in Presence of Artificial Saliva

Fabio Alemanno [1], Veronica Peretti [2], Angela Tortora [1] and Silvia Spriano [2,*]

[1] Ducom Instruments Europe B.V., Zernikepark 6, 9747AN Groningen, The Netherlands; f.alemanno91@gmail.com (F.A.); angela.tortora@ducom.com (A.T.)
[2] Politecnico di Torino, Corso Duca degli Abruzzi, 24-10129 Torino, Italy; veronica.peretti88@gmail.com
* Correspondence: silvia.spriano@polito.it

Received: 30 July 2020; Accepted: 26 August 2020; Published: 31 August 2020

Abstract: Abutment is the transmucosal component in a dental implant system and its eventual appearance has a major impact on aesthetics: use of zirconia abutments can be greatly advantageous in avoiding this problem. Both in the case of one and two-piece zirconia abutments, a critical issue is severe wear between the zirconia and titanium components. High friction at this interface can induce loosening of the abutment connection, production of titanium wear debris, and finally, peri-implant gingivitis, gingival discoloration, or marginal bone adsorption can occur. As in vivo wear measurements are highly complex and time-consuming, wear analysis is usually performed in simulators in the presence of artificial saliva. Different commercial products and recipes for artificial saliva are available and the effects of the different mixtures on the tribological behaviour is not widely explored. The specific purpose of this research was to compare two types of artificial saliva as a lubricant in titanium–zirconia contact by using the ball on disc test as a standard tribological test for materials characterisation. Moreover, a new methodology is suggested by using electrokinetic zeta potential titration and contact angle measurements to investigate the chemical stability at the titanium–lubricant interface. This investigation is of relevance both in the case of using zirconia abutments and artificial saliva against chronic dry mouth. Results suggest that an artificial saliva containing organic corrosion inhibitors is able to be firmly mechanically and chemically adsorb on the surface of the Ti c.p. or Ti6Al4V alloy and form a protective film with high wettability. This type of artificial saliva can significantly reduce the friction coefficient and wear of both the titanium and zirconia surfaces. The use of this type of artificial saliva in standard wear tests has to be carefully considered because the wear resistance of the materials can be overestimated while it can be useful in some specific clinical applications. When saliva is free from organic corrosion inhibitors, wear occurs with a galling mechanism. The occurrence of a super-hydrophilic saliva film that is not firmly adsorbed on the surface is not efficient in order to reduce wear. The results give both suggestions about the experimental conditions for lab testing and in vivo performance of components of dental implants when artificial saliva is used.

Keywords: artificial saliva; lubricant; zirconia; titanium alloys; wear

1. Introduction

The term biotribology comprises tribological phenomena occurring both in natural tissues and organs of living organisms (such as wear of skin or sliding of eyelids over eye) and after implantation of an artificial device (wear of orthopaedic and dental implants) [1]. This research focused on the last case, and deals with the tribology of materials used for dental implants with a focus on the interface between titanium and zirconia.

Wear of natural or artificial teeth mainly results from the closed phase of chewing, with food–tooth contact, or eventually from thegosis (sliding teeth into lateral positions) and bruxism (grinding teeth without the presence of food) when direct tooth–tooth contact occurs. Furthermore, nowadays, chemical effects play an increasingly important role in tooth wear, mainly as a result of the large consumption of acid drinks (erosion). Additionally, tooth wear can also result from tooth cleaning or pipe smoking. If moderate tooth wear has significant clinical consequences (aesthetically and functionally), excessive wear can result in unacceptable damage of the occluding surfaces, alteration of the functional path of masticatory movement, dentine hypersensitivity, and pulpal pathology [2]. Biotribology must include both standard tests for materials characterization and specific tests for the simulation of the in vivo environment and mechanical loading [3]. This research focused on the first topic, but with a focus on the selection of the lubricant, which is, in this case, artificial saliva.

Saliva is the lubricant in the human mouth and consists of approximately 98% water plus a variety of electrolytes and proteins [4]. Salivary proteins can be selectively adsorbed onto solid substrates as well as mucosa membranes exposed to an oral environment, forming a pellicle within a few seconds. Saliva is a source of inorganic ions necessary for remineralization and self-repair capacity because it supplies calcium and phosphate ions. The role of the salivary pellicle is also thought to protect the underlying tooth surface from acid attack because it is a buffer to acids introduced into the mouth [5]. An important function of saliva is to form a boundary lubrication system and to act as a lubricant between hard (tooth) and soft (mucosal) tissues or in dental implants, decreasing wear and reducing friction [5]. It has been widely reported that a salivary film is formed by a layer-by-layer adsorption of salivary proteins and then has a heterogeneous structure consisting of a thin and dense inner layer, able to decrease the friction and wear of teeth, and a thick, highly hydrated and viscoelastic outer layer [6]. The lubricating properties of saliva are presumably related to the strong adhesion strength of the inner salivary film on the tooth surface, which depend on surface roughness, free energy, and the presence of positive or negative surface charges [5]. Harvey et al. observed that the friction coefficient on a hydrophobic substrate was almost an order of magnitude higher than that on a hydrophilic one because of the lower adhesion strength of the salivary film [7]. On the other side, hydrophobic materials in the oral cavity might be more easily cleaned from adsorbed salivary films.

As the transmucosal component in the implant system, the abutment is exposed to the oral cavity and its eventual appearance has a major impact on the aesthetics: using zirconia abutments, aesthetics can be indeed improved. Zirconia abutments are available in two basic designs: one-piece abutment (all parts are made of zirconia in a unit) or two-piece abutment (a secondary metallic component is used as connecting part). Several studies have verified that the fracture strength of two-piece abutments is higher than that of the one-piece design [8,9]. In both cases, the critical issue is severe wear between the zirconia and titanium inducing loosening of the abutment connection and production of titanium wear debris, which can result in peri-implant gingivitis, gingival discoloration, or marginal bone adsorption [10].

A comprehensive description of the tribocorrosion behaviour of different materials coupled in a dental implant can be found in a paper by Sikora et al. [11]. The specific aim of this research was to compare two types of artificial saliva as a lubricant in titanium–zirconia contact and both mechanical (through tribological test ball on disc) and chemical stability (through zeta potential titration curves) of the lubricant film were tested. This investigation is of relevance both for the wear testing of materials and in the case of the clinical use of artificial saliva against chronic dry mouth.

2. Materials and Methods

The discs used for tribological tests were made of commercially pure titanium (ASTM B348, Gr2, Titanium Consulting and Trading, Florence, Italy) and Ti6Al4V alloy (ASTM B348, Gr5, Titanium Consulting and Trading, Florence, Italy) and obtained from cylindrical bars (10 mm in diameter and 2 mm in thickness). Prior to the tests, the discs were polished using abrasive paper with different grit values (320–2000) and then ultrasonically cleaned in ethanol for 10 min twice and dried with compressed air. The balls used were made of zirconium oxide (ZrO_2) partially stabilized with yttrium

(Y-TZP; Ceratec Technical Ceramics BV, Geldermalsen, The Netherlands; composition: 94.7 wt.% ZrO_2, 5.3 wt.% Y_2O_3; 6 mm in diameter).

The aim of the research was to compare artificial saliva of different compositions. To do so, two commercial products were used in the form of liquids: Biotène (Biotène® Moisturizing Spray) and BioXtra (Biopharm). Both fluids contain water, xylitol, sodium benzoate, paraben (propylparaben and methylparaben or sodium propylparaben), and sodium saccharin. Moreover, Biotène contains organic corrosion inhibitors: polyethylene glycol 60 hydrogenated castor oil, cetylpyridinium chloride, and the polymer vp/va copolymer. Finally, Biotène contains glycerin, xanthan gum, limonene (pH = 6.06), while BioXtra contains hydrogenated starch hydrolysate, sorbitol, hydroxyethylcellulose, sodium monofluorophosphate, potassium chloride, sodium chloride, magnesium chloride, dipotassium phosphate, calcium chloride, colostrum whey, lactoperoxidase, citric acid, and potassium sorbate (pH = 7.33). No organic corrosion inhibitor is present in BioXtra.

Tribological tests were performed by using a high frequency reciprocating rig instrument (TR-282, Ducom Instruments Pvt. Ltd., Bengaluru, India) with the following parameters: Load = 2 N; Frequency = 30 Hz; Stroke length = 1 mm; Velocity = 60 mm/s; Temperature = 37 °C; Duration = 60 min; Lubrication volume = 2 mL. The lubricant (artificial saliva) was measured with a syringe and poured into the reservoir of the disc holder, completely covering the sample. The test was repeated on three different discs (and balls) for each pair of materials and saliva lubricant: the set of experiments allowed us to independently test the different variables. The aim was to independently compare the effect of the different saliva on the same substrate (Ti c.p. or Ti6Al4V) and of the same type of saliva on different substrates. The results were statistically analysed by using the Student's t-test.

Two-dimensional images of the discs and balls were acquired both before and after each test using optical microscopes (AmScope, ME300 Series and Olympus Vanox-T, Leica Microsystems B.V., Amsterdam, The Netherlands, respectively, for observing the discs and balls) in order to measure the length and width of the wear scars (measurements performed by ImageJ software (version 1.52). The samples after tribological tests were rinsed and wiped with ethanol before imaging.

Adsorption of saliva before zeta potential titration and contact angle measurements were performed by placing discs in a 12-well container and each one was covered with 0.5 mL of saliva. The adsorption phase lasted 1 h at 37 °C in the incubator. Subsequently, the samples were left to air dry for two days.

An electrokinetic analyser (SurPASS, Anton Paar, Graz, Austria) was employed for zeta potential titration vs. pH on the discs after saliva absorption. The surface zeta potential was determined in function of pH in a 0.001 M KCl electrolyte solution, varying the solution pH by the addition of 0.05 M HCl or 0.05 M NaOH through the instrument automatic titration unit. The acid and alkaline sides of the curve were obtained in two different steps on two different couples of samples.

Surface wettability by water was determined on the discs after the zeta potential tests through contact angle measurements (sessile drop method, DSA-100, KRÜSS GmbH, Hamburg, Germany). Briefly, a drop (5 µL) of ultrapure water was deposited on the material surface and the angle formed with the surface measured by the instrument software.

3. Results and Discussion

The coefficient of friction (COF) was measured during the tribological tests of zirconia balls on Ti c.p. or Ti alloy discs under lubrication with different artificial saliva: Biotène (containing organic corrosion inhibitors) and BioXtra (without organic corrosion inhibitors). Figure 1 shows the average friction coefficient as a function of the sliding distance for the four tested material/saliva combinations.

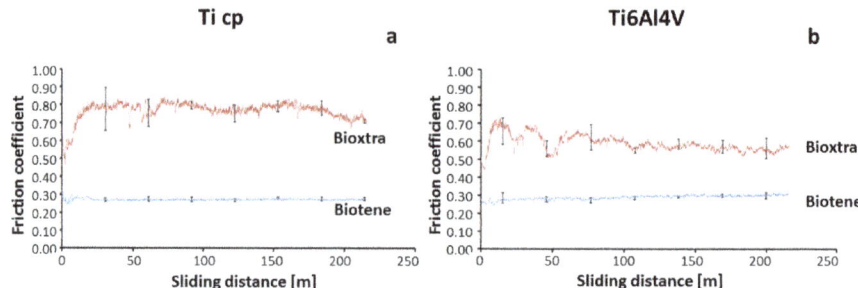

Figure 1. Average COF over sliding distance of Ti c.p. (**a**) and Ti6Al4V alloy (**b**) in the presence of Biotène (blue) or Bioxtra (red).

Compared to the BioXtra saliva, Biotène decreased the friction coefficient on both substrates with good lubrication action, as can be seen from the values shown in Table 1: the coefficients of friction of both materials were much lower in the presence of Biotène and they do not differ significantly from each other ($p > 0.05$, Student's t-test). The difference between the average friction coefficient of Ti c.p. in the presence of Biotène or BioXtra as a lubricant was statistically significant ($p < 0.01$); the same observation can be made in the case of Ti6Al4V. If we consider BioXtra as a lubricant and we compare Ti c.p. with Ti6Al4V, the difference of the average friction coefficients is again statistically significant ($p < 0.01$). Considering Biotène as a lubricant, the difference between the average friction coefficient of the two substrates was statistically less evident ($p < 0.05$). Furthermore, Biotène showed, by far, lower variability in the friction coefficient. In fact, there was a clear difference in the curve trend: samples lubricated with BioXtra had an initial region with gradual increase of COF, which means a "running-in period" that was much more extensive than those lubricated with Biotène. In this region, an adjustment of the two contacting surfaces occurred by crushing and smearing of the asperities. Then, a second region can be identified, during which COF remains fairly stable in the presence of Biotène, while the COF of the samples lubricated with BioXtra exhibited strong oscillations. After the running-in period, these oscillations may be attributed to the build-up and accumulation of third-body particles in the contact region.

Table 1. Average friction coefficients and standard deviations of all the tested samples.

Sample	Lubricant	Test 1	Test 2	Test 3	Average	Standard Deviation
Ti c.p.	Biotène	0.27	0.27	0.28	0.27	0.005
	Bioxtra	0.73	0.76	0.80	0.77	0.033
Ti6Al4V	Biotène	0.29	0.30	0.28	0.29	0.011
	Bioxtra	0.58	0.59	0.60	0.59	0.009

A picture of the wear scar was acquired from each disc at the end of each test and a representative set of images is reported in Figure 2 for each of the tested materials.

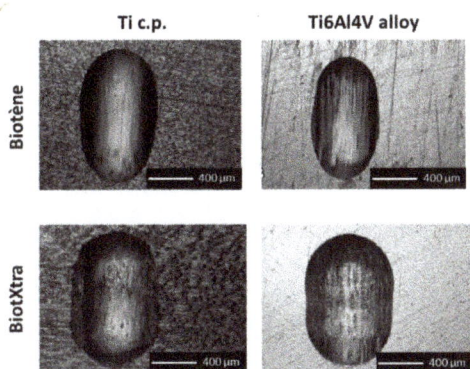

Figure 2. Images of the wear scars on Ti c.p and Ti6Al4V after the tests with Biotène and BioXtra.

In the presence of Biotène (Figure 2), the shape of the scars on both materials was very similar and was almost a perfect ellipse. The edges were regular and well defined. The surface on the bottom of the scars was uniform and almost free of debris. Instead, in the presence of BioXtra (Figure 2), the scars on the two materials resulted in a more irregular elliptical shape. The edges were more jagged and the surface on the bottom of the scars was rough with a considerable presence of debris. In both cases, it can be seen that the grooves on the bottom of the scars showed an oriented texture aligned to the sliding direction. This can be explained by third-body abrasion processes caused by the formation of metal debris due to the ultra-hardness of zirconia.

The area of wear scars in the presence of Biotène was smaller than that in the presence of BioXtra (Figure 3); in particular, that on the titanium alloy was the lowest of them all, according to the higher mechanical resistance of the Ti6Al4V alloy with respect to Ti c.p.

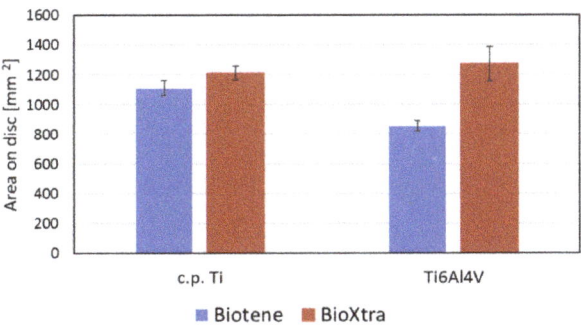

Figure 3. Average areas of the wear scars on the discs.

An image of the wear scars on the zirconia balls was taken at the end of each test and a representative picture is reported below for each tribo-pair in Figure 4.

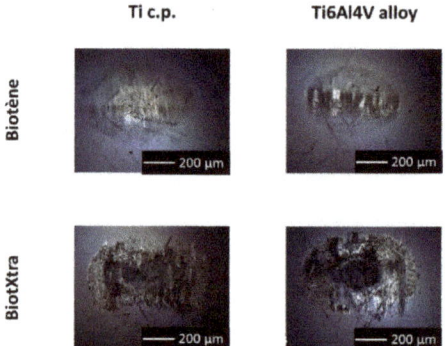

Figure 4. Images of a zirconia balls used against Ti cp and Ti6Al4V discs after the tests with Biotène and BioXtra.

It can be immediately noticed that the wear tracks on the zirconia balls tested in the presence of Biotène had a similar aspect and the same for those tested with BioXtra. In all cases, the wear scars formed on zirconia balls had an elliptical shape with a darker and more worn-out central area. The occurrence of a strong adhesive interaction between the tribo-pairs lubricated with BioXtra was supported by the higher amount of material transfer on the zirconia balls.

The mean wear scar diameter (MWSD) was calculated by averaging the two main diameters of wear scars on the zirconia ball (Figure 5). The MWSDs registered on Ti c.p. were not statistically different if the scars obtained in the tests with the two different lubricants are compared; no statistical difference also existed between the two substrates tested by using BioXtra as the lubricant. On the other hand, this difference between the MWSDs was statistically significant on Ti6Al4V and if the two substrates tested with Biotène are compared ($p < 0.01$)

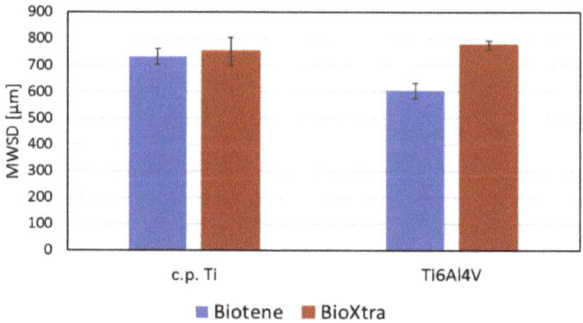

Figure 5. Average mean wear scar diameters on the zirconia balls.

At the end of the tests, the amount of metallic debris dispersed in the artificial saliva used as a lubricant was qualitatively observed (data not reported); once again, the difference between the two salivas was evident with a greater quantity of residues visible at the end of the tests carried out with BioXtra.

It therefore appears that Biotène decreases the wear between the tested materials; this is due first of all to the lower friction coefficient than that in case of BioXtra. Furthermore, the presence in this artificial saliva of organic corrosion inhibitors (surfactants PEG-60 hydrogenated castor oil, cetylpyridinium chloride, and the polymer vp/va copolymer) can justify the lower tendency to wear. In fact, the mechanism of action of organic corrosion inhibitors is based on the adsorption on the surface to form a protective film that protects it against deteriorating [12]. In the presence of BioXtra, a strong adhesive interaction has occurred between zirconia and Ti c.p. or the Ti alloy. Hence, the predominant wear mechanism is adhesive galling. It is likely that the surfactants, absent in this artificial saliva and present in Biotène, provide some protection to titanium surfaces from galling.

In this research, in addition to the mechanical tests, the chemical stability of the lubricant film on the titanium surfaces was tested by means of zeta potential titration curves. This is of interest, considering that different pH values can be present in the mouth and that wettability of the lubricant film and chemical interaction at the lubricant–surface interface are critical in determining the biotribological behaviour. The graphs below (Figure 6) show the zeta potential titration curves of the Ti c.p. and Ti alloy discs after adsorption of the two artificial salivas.

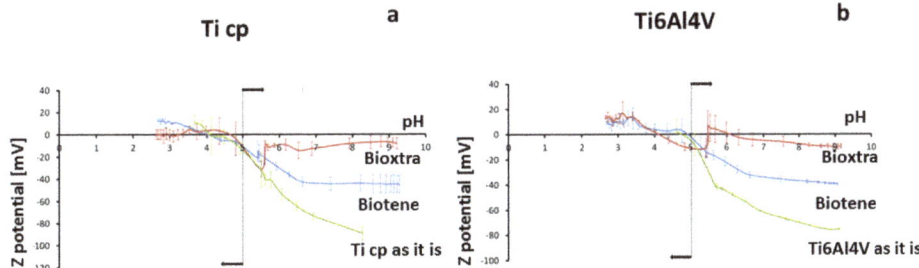

Figure 6. Zeta potential titration curves of Ti c.p. (**a**) and Ti6Al4V alloy (**b**) as received (green curves), after adsorption of Biotène (blue curves) and of BioXtra (red curves). The arrows mark the pH used at the beginning of the measurements and the acidic (on the left) and alkaline (on the right) ranges of the titration.

Comparing the curves of the samples pre-adsorbed with Biotène with the curves of the as-received substrates (i.e., tested without saliva pre-adsorption), we note that the curves obtained on the samples after adsorption were significantly different from the curves of the as-received materials. This means that the adsorption of some compounds from saliva occurred. The slope of the titration curves after adsorption was significantly lower than that of the as-received materials, which means that hydrophilicity of the surfaces after adsorption is higher than that of the bare metals. There was a plateau at pH higher than 6.5 due to functional groups (such as, for instance, OH groups) exposed on the surface by the adsorbed compounds, meaning that the functional groups on the surfaces have a specific acidic strength and are all completely dissociated when the pH is higher than 6.5, that means that in physiological condition there is a net negative surface charge. According to their dissociation only in the basic range, these groups do not act as a strong acid and they do not significantly affect the isoelectric point on the curve with respect to the Ti as it is (without adsorption of saliva). On the other hand, when strong acidic or alkaline functional groups are present on a surface, they are respectively de/protonated at lower/higher pH values than 7 and both the isoelectric point and the onset of the plateau are shifted to the left/right side.

Observing the curves related to the adsorption of BioXtra, it can be said that the curves obtained after adsorption were significantly different from the curves of the as-received materials, which means that the adsorption of saliva compounds also occurred in this case. The adsorbed layer had zeta potential values not far from 0 across the explored range: this behaviour is typical of a super-hydrophilic surface that does not adsorb either OH^- or H_3O^+ ions instead of water molecules, even if the solution is strongly basic or acidic. On the other hand, the curve was not stable with a high standard deviation at some points and fluctuating trend, according to the presence of an un-stable surface layer. In addition, the step registered at pH 5.5 (that is, the first point of the two titration curves in the basic and acidic range) can be related to some instability, or not perfect reproducibility of the adsorbed layer.

These results can be used for interpreting the results obtained from wear tests. Compared to Bioxtra, Biotène decreased the friction coefficient and wear of both substrates because it allows for the formation of a chemically stable adsorbed layer with a good hydrophilic behaviour, acting as a good lubricant. On the other hand, the layer formed by BioXtra is not chemically stable and it is not able to act as a good lubricant, even if it has a super-hydrophilic behaviour.

This hypothesis was confirmed by contact angle measurements performed with a water drop on the discs after zeta potential tests, as reported in Figure 7. A consistent decrease in the contact angle with respect to the as-received Ti surfaces (without any adsorption of saliva) was measured on the sample pre-adsorbed with Biotène and used for the zeta potential test, which was in agreement with the zeta potential titration curve where a higher wettability was detected after adsorption of Biotène and confirmed that this saliva was still firmly adsorbed after the electrokinetic zeta potential test. On the other hand, the contact angle was almost unchanged with respect to the as-received titanium surfaces (without any adsorption of saliva) on the sample pre-adsorbed with BioXtra and used for the zeta potential test, confirming that the saliva film was almost completely removed after the electrokinetic zeta potential test and was not stably adsorbed on the titanium surfaces.

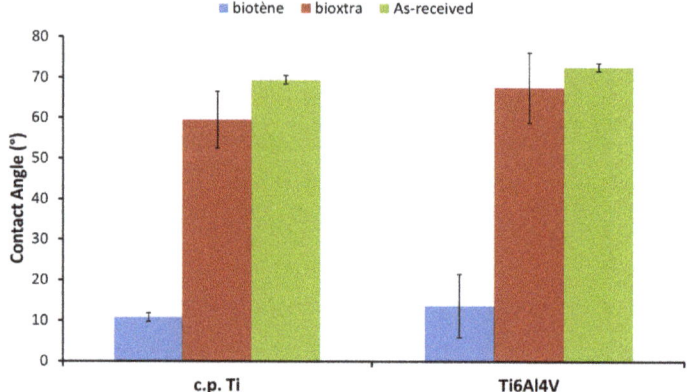

Figure 7. Contact angle measurements of water on Ti c.p. and Ti6Al4V alloy as received and after adsorption of the Biotene and BioXtra artificial salivas.

The obtained results are of interest because they set the relevance of the composition of the selected artificial saliva on the wear resistance of dental materials. Even if this topic has been already partially explored in the literature [13–15], the new evidence derived from this research are here discussed. Friction coefficient and wear were significantly reduced in the presence of organic corrosion inhibitors able to make a chemically and mechanically stable hydrophilic adsorbed layer. First, this means that researchers planning tribological tests have to carefully select the artificial saliva used as a lubricant in order to avoid overestimating the wear resistance of the tested surfaces. Second, it means that patients who need to use artificial saliva and have dental implants have to carefully select the right one. Finally, the use of zeta potential titration measurements is proposed here as a useful experimental tool for obtaining more information for a better understanding of the tribological tests: presence/absence of an adsorbed layer of lubricant on a tribological surface, its chemical and mechanical stability when significant pH changes occur and a flux of liquid is applied on the surface, the presence of functional groups exposed, and hydrophilicity of the surface after adsorption of the lubricant. The chemical instability of the adsorbed layer of saliva at pH 5.5 in the case of Bioxtra may be relevant in physiological conditions considering that acidic pH can be easily reached in the oral environment.

On the basis of this research, other investigations will be performed in order to overcome the limitations of the tests previously described including the investigation of the Ti6Al4V-ELI alloy (Gr23) as a substrate because it is currently widely employed in dental applications, better investigation of the adsorbed film thickness, comparison of artificial saliva with different additives (such as the antibacterial ones), chemical analysis of the adsorbed compounds, different geometry of the samples, and tribological tests (such as flat-on-flat tests) that better simulate the tribological contact in dental implants.

4. Conclusions

Results suggest that the use of an artificial saliva containing organic corrosion inhibitors (such as Biotène) can significantly reduce the friction coefficient and wear both of titanium and zirconia surfaces. This is because it is able to be firmly adsorbed on the surface of the Ti c.p. or Ti6Al4V alloy and form a protective film with high wettability. In the absence of this film (such as in the case of BioXtra), much more severe wear occurs with a galling mechanism. In this last case, the occurrence of a super-hydrophilic saliva film that is not firmly adsorbed on the surface (mainly in an acidic environment) is not efficient in order to avoid severe wear.

From the methodological standpoint, the use of electrokinetic zeta potential and contact angle measurements together with standard tribological tests has shown to be useful for a better understanding of the lubricant action of different artificial saliva on different materials.

This research can be the basis for larger investigations of the tribological behaviour of dental materials and implants in the presence of lubricants with different compositions.

Author Contributions: Conceptualization, F.A., A.T. and S.S.; methodology, F.A. and A.T.; investigation, F.A., A.T. and V.P.; writing—original draft preparation, S.S.; writing—review and editing, S.S.; supervision, S.S. All authors have read and agreed to the published version of the manuscript.

Funding: This research received no external funding.

Acknowledgments: The authors would like to acknowledge Ducom Instruments Europe B.V. for the friction and wear measurements.

Conflicts of Interest: The authors declare no conflict of interest.

References

1. Zhou, Z.R.; Jin, Z.M. Biotribology: Recent progresses and future perspectives. *Biosurf. Biotrib.* **2015**, *1*, 3–24. [CrossRef]
2. Zheng, J.; Zhou, Z.-R. Oral Tribology. In *Encyclopedia of Tribology*; Wang, Q.J., Chung, Y.-W., Eds.; Springer: Boston, MA, USA, 2013.
3. Branco, A.C.; Moreira, V.; Reis, J.A.; Figueiredo-Pina, C.G.; Serro, A.P. Influence of contact configuration and lubricating conditions on the microtriboactivity of the zirconia-Ti6Al4V pair used in dental applications. *J. Mech. Behav. Biomed. Mat.* **2019**, *91*, 164–173. [CrossRef] [PubMed]
4. De Almeida, P.V.; Gregio, A.; Machado, M.; de Lima, A.; Azevedo, L.R. Saliva composition and functions: A comprehensive review. *J. Contemp. Dent. Pract.* **2008**, *9*, 72–80.
5. Heller, D.; Helmerhorst, E.J.; Oppenheim, F.G. Saliva and serum protein exchange at the tooth enamel surface. *J. Dent. Res.* **2017**, *96*, 437–443. [CrossRef] [PubMed]
6. Macakova, L.; Yakubov, G.; Plunkett, M.A.; Stokes, J.R. Influence of ionic strength changes on the structure of pre-adsorbed salivary films. A response of a natural multi-component layer. *Coll. Surf. B Biointerfaces* **2010**, *77*, 31–39. [CrossRef] [PubMed]
7. Harvey, N.M.; Carpenter, G.H.; Proctor, G.B.; Klein, J. Normal and frictional interactions of purified human statherin adsorbed on molecularly-smooth solid substrata. *Biofouling* **2011**, *27*, 823–835. [CrossRef] [PubMed]
8. Stimmelmayr, M.; Sagerer, S.; Erdelt, K.; Beuer, F. In vitro fatigue and fracture strength testing of one-piece zirconia implant abutments and zirconia implant abutments connected to titanium cores. *Int. J. Oral Maxillofac. Implant.* **2013**, *28*, 488–493. [CrossRef] [PubMed]
9. Gehrke, P.; Johannson, D.; Fischer, C.; Stawarczyk, B.; Beuer, F. In vitro fatigue and fracture resistance of one- and two-piece CAD/CAM zirconia implant abutments. *Int. J. Oral Maxillofac. Implant.* **2014**, *30*, 546–554. [CrossRef] [PubMed]
10. Mizumoto, R.M.; Malamis, D.; Mascarenhas, F.; Tatakis, D.N.; Lee, D.J. Titanium implant wear from a zirconia custom abutment: A. clinical report. *J. Pros. Dent.* **2019**, in press.

11. Sikora, C.L.; Alfaro, M.F.; Yuan, J.C.-C.; Barao, V.A.; Sukotjo, C.; Mathew, M.T. Wear and corrosion interactions at the titanium/zirconia interface: Dental implant application. *J. Prosthodont.* **2018**, *27*, 842–852. [CrossRef] [PubMed]
12. Brycki, B.E.; Kowalczyk, I.H.; Szulc, A.; Kaczerewska, O.; Pakiet, M. Organic Corrosion Inhibitors. In *Corrosion Inhibitors, Principles and Recent Applications*; Aliofkhazraei, M., Ed.; Tarbiat Modares University: Tehran, Iran, 2017.
13. Vilhena, L.; Oppong, G.; Ramalho, A. Tribocorrosion of different biomaterials under reciprocating sliding conditions in artificial saliva. *Lubr. Sci.* **2019**, *31*, 364–380. [CrossRef]
14. Li, C.; Liang, R.; Ren, J.; Wang, J.; Xun, Y.; Meng, H.; Sun, S. Comparative study on friction properties of different dental restorative materials against natural tooth enamel and dentin. *J. Mat. Res.* **2016**, *30*, 489–495.
15. Borrás, A.D.; Buch, A.; Cardete, A.R.; Navarro-Laboulais, J.; Muñoz, A.I. Chemo-mechanical effects on the tribocorrosion behavior of titanium/ceramic dental implant pairs in artificial saliva. *Wear* **2019**, *426–427*, 162–170. [CrossRef]

© 2020 by the authors. Licensee MDPI, Basel, Switzerland. This article is an open access article distributed under the terms and conditions of the Creative Commons Attribution (CC BY) license (http://creativecommons.org/licenses/by/4.0/).

Article

Influence of Conditions for Production and Thermo-Chemical Treatment of Al₂O₃ Coatings on Wettability and Energy State of Their Surface

Mateusz Niedźwiedź, Władysław Skoneczny * and Marek Bara

Institute of Materials Engineering, Faculty of Science and Technology, University of Silesia in Katowice, 40-007 Katowice, Poland; mateusz.niedzwiedz@us.edu.pl (M.N.); marek.bara@us.edu.pl (M.B.)
* Correspondence: wladyslaw.skoneczny@us.edu.pl; Tel.: +48-32-368563

Received: 26 June 2020; Accepted: 13 July 2020; Published: 15 July 2020

Abstract: This article presents the influence of the anodizing parameters and thermo-chemical treatment of Al_2O_3 coatings made on aluminum alloy EN AW-5251 on the surface free energy. The oxide coating was produced by DC (Direct Current) anodizing in a ternary electrolyte. The thermo-chemical treatment of the oxide coatings was carried out using distilled water, sodium dichromate and sodium sulphate. Micrographs of the surface of the Al_2O_3 coatings were characterized using a scanning microscope (SEM). The chemical composition of the oxide coatings was identified using EDS (Energy Dispersive X-ray Spectroscopy) microanalysis. Surface free energy (SFE) calculations were performed by the Owens–Wendt method, based on wetting angle measurements made using the sessile drop technique. The highest value of surface free energy for the only anodized coatings was 46.57 mJ/m^2, and the lowest was 37.66 mJ/m^2. The contact angle measurement with glycerine was 98.06° ± 2.62°, suggesting a hydrophobic surface. The thermo-chemical treatment of the oxide coatings for most samples contributed to a significant increase in SFE, while reducing the contact angle with water. The highest value of surface free energy for the coatings after thermo-chemical treatment was 77.94 mJ/m^2, while the lowest was 34.98 mJ/m^2. Taking into account the contact angle measurement with glycerine, it was possible to obtain hydrophobic layers with the highest angle of 109.82° ± 4.79° for the sample after thermal treatment in sodium sulphate.

Keywords: aluminum oxide layers; surface morphology; thermo-chemical treatment

1. Introduction

Nowadays, when a common criterion for material selection is its weight to strength ratio, and when great importance is attached to environmentally friendly solutions, aluminum has become very widely used in material engineering [1]. The most important advantages of aluminum are undoubtedly its low weight in relation to strength, high electrical and thermal conductivity, ease of forming in all machining processes such as rolling and extrusion, considerable corrosion resistance and almost 100% recyclability without losing its properties [2–4]. In its pure form, aluminum, in addition to the number of advantages, also has significant disadvantages, which are its low melting point and low absolute strength [5]. In order to minimize these disadvantages, aluminum is combined with elements such as copper, silicon or magnesium to form durable alloys [6]. The industries that most often use aluminum and its alloys are the automotive, aviation and shipbuilding industries [7–10].

In order to better protect aluminum alloys against mechanical damage and corrosion, the process of so-called anodizing (oxidation) is used [11,12]. Anodizing aluminum and its alloys is an electrochemical process consisting of creating an oxide coating on the surface of an alloy, usually with a thickness of a few to several dozen micrometers [13]. The anodized coating is many times harder than pure aluminum or aluminum alloys (1700 to 2000 HV Al_2O_3 hardness at about 30 HV for pure aluminum and 200 HV for PA9 alloy) [14,15]. During the electrochemical process, the formation of the oxide coating occurs at the expense of substrate loss. This results in very good adhesion of the coating to the aluminum substrate. The most commonly used type of anodizing is hard DC (Direct Current) anodizing. During the process, as the coating thickness increases, the anodizing voltage increases [16]. The beginning of the DC anodizing process is characterized by a rapid increase in voltage, necessary to break through the compact and thin (0.01–0.1 μm thick) natural barrier layer, also called the barrier layer [17]. In the next stage of the process, the voltage decreases. When the minimum value is reached, the barrier layer is rebuilt at the oxide–electrolyte interface. The voltage increases again over time due to the formation of pores and their further deepening as the thickness of the Al_2O_3 coating increases, up to a maximum thickness of approximately 150 μm [18,19]. The oxide coating produced during anodizing ideally reaches the form of regular pores arranged in the shape of a hexagonal system, along with a thin barrier film in direct contact with the aluminum substrate. The oxide pore sizes mainly depend on the anodizing method used, but also on the process parameters. A high temperature and long anodizing time increase the solubility of the oxide, while creating a porous coating with poorer protective properties [20,21].

For better corrosion protection and sealing of the absorbent oxide coating while maintaining its advantages, thermo-chemical treatment is used [22]. As a result of processing in the Al_2O_3 coating, aluminum oxide is transformed into hydrated forms such as γ-Al (OH)$_3$ hydrargillite and γ-AlOOH boehmite. The transformation occurs due to swelling of the walls of the Al_2O_3 cells owing to their hydration, as a result of which the pores are closed, and a smooth surface is created [23]. During thermo-chemical treatment, the process of producing a pseudo-boehmite sublayer on the surface may also occur; this happens most often when the process is short (several minutes), but also at high temperatures [24].

Researchers have investigated the thermo-chemical treatment of oxide coatings, but the research lacks reference to the impact of this process on the wettability of the Al_2O_3 layers [25,26]. The porosity of oxide coatings has a significant impact on their energy state, and thus the properties enabling the repulsion (hydrophobicity) and attraction (hydrophilicity) of water molecules [27]. That is why studies on the wettability of oxide coatings produced as a result of anodizing and thermo-chemical treatment are so important. Nowadays, the energy state is a very important feature of materials. Materials with hydrophobic properties, i.e., those exhibiting low wettability, repelling water particles, are widely used in surface engineering as self-cleaning or anti-icing materials [28,29]. Materials with high wettability are also in high demand, or, in other words, attracting water particles (hydrophilic); they are used in photovoltaic panels, biomaterials and wherever it is important to reduce friction [30–32]. A surface is considered hydrophobic when the water contact angle is greater than 90°. The conventional name for surfaces reaching contact angles over 150° is ultrahydrophobicity. The opposite of hydrophobicity is hydrophilicity and occurs when the contact angle on the surface is less than 90° [33].

Many authors have carried out research on the modification of Al_2O_3 layers by thermo-chemical treatment. In [34], 6061 aluminum alloy was anodized at a constant voltage in an electrolyte at the temperature of 273 K. A 15% H_2SO_4 solution was used as the electrolyte, the anodizing time was 60 min, and a voltage of 15–30 V was used. Then, each of the produced layers were modified by immersion in stearic acid at the temperature of 353 K for 45 min, successively in ethanol at the temperature of 343 K. The whole process was completed by drying in an oven at the temperature of 253 K. The authors of the work managed to obtain hydrophobic surfaces; the contact angles were 138°–152° depending on the anodizing voltage used. In subsequent studies, researchers conducted studies using two-stage anodizing on 1050 aluminium alloy at the temperature of 273 K. As the electrolyte, 10.5% and 1% solutions of H_3PO_4 were used; the process was carried out using the constant voltage method at voltages in the range of 160–195 V for 60 min in Stage 1 and 15–300 min in Stage 2. After Stage 1 of anodizing, the layers were immersed in a solution of 6% H_3PO_4, 1.8% CrO_3 and deionized water at the temperature of 333 K. Wet chemical etching was carried out in a 5% solution of H_3PO_4 at the temperature of 333 K for 60–150 min. In the next step, the layers were sonicated in acetone for 10 min, then dried in an oven at the temperature of 343 K for 360 min and modified by immersion in a 5% $C_{12}H_{24}O_2$ solution for 90 min. The last stage of layer modification was spraying with silane and dried at the temperature of 413 K for 30 min. Thanks to the use of lauric acid, the contact angles were higher by 5°–30°, while thanks to silane spraying, a hydrophobic coating with a contact angle of 146° was obtained [35]. In the next cited article, researchers also dealt with constant voltage anodizing using a $C_2H_2O_4$ electrolyte. The anodizing time was 420 min at the voltage of 40 V. The layers were modified by immersion in an 0.3% polytetrafluoroethylene solution and then cured at room temperature. In the next stage, the samples were immersed in a solution of 1.8% H_2CrO_4 and 6% H_3PO_4 at the temperature of 338 K for 300 min. Contact angles of 120° before immersion and 160° after immersion in a polytetrafluoroethylene solution were obtained [36]. Another article that can be cited is research based on two-stage constant voltage anodizing. Anodizing was carried out at the temperature of 275 K in a 1% H_3PO_4 solution. A voltage of 194 V was applied for 60 min in the 1st stage and 300 min in the 2nd stage. After completing Stage 1, the layers were immersed in a 6% H_3PO_4 solution and a 2% H_2CrO_4 solution at the temperature of 323 K for 40 min. After anodizing was completed, the layers were immersion modified in deionized H_2O at the temperature of 373 K for 1 min. This process mainly involved preparing the substrate for subsequent treatments. The samples were then dried at the temperature of 323 K and finally treated with hexamethyldisilane vapor (HMDS) for 240 and 540 min. As a result of the performed modifications, contact angles of 139°–153° were obtained (depending on the number and duration of HMDS cycles) [37].

In all the cited publications, the researchers focused mainly on the modification of oxide coatings produced as a result of constant voltage anodizing. In the next publication, the coatings were prepared using the DC method for 120 min at a current density of 1–2 A/dm^2. The process was carried out in a 5% H_3PO_4 solution at a temperature of 283–293 K. Some of the samples were plasma treated and all of them were immersed in trichloro octadecylsilane-hexane for 120 min as well as being washed with hexane and oven dried at the temperature of 333 K. Wetting angles of 152° (without plasma treatment) and 157° (after plasma treatment) were obtained [38].

After analyzing the literature, it was found that most of the world research on the wettability of Al_2O_3 coatings is mainly based on the modification of the coating by applying additional layers to their surfaces. There is a lack of research in this area regarding the impact of the anodizing parameters before and after applying thermo-chemical treatment and the impact of the compounds used for this treatment on the contact angles, and thus the energy state of the surface of the coatings, which makes our research innovative. The studies presented below focused mainly on the impact of the anodizing parameters (current density, electrolyte temperature, anodizing time) and the impact of the thermo-chemical treatment carried out with various compounds (water, sodium dichromate, sodium sulphate) on the contact angles of oxide coatings. Measurement of the contact angles allowed the energy state of the layers to be determined by calculating the surface free energy.

2. Materials and Methods

2.1. Research Material

The research material was Al_2O_3 coatings made on aluminum alloy EN AW-5251 by means of hard DC anodizing. The aluminum alloy used has a high magnesium content, which contributes to such properties as good ductility, high corrosion resistance and high weldability. This alloy is most often used in the construction of aircraft and marine structures.

The samples for producing the oxide coatings had a surface area of 0.1 dm^2. Before proceeding with the anodizing process, each sample was etched in a 5% KOH solution for 20 min, followed by a whitening process in a 10% HNO_3 solution for 5 min. The temperature of the used solutions was 296 K. At the end of each process, the samples were rinsed in distilled water. The coatings were produced in the DC anodizing process using a GPR-25H30D power supply. The process was carried out in variable production conditions (current density, anodizing time), at a variable electrolyte temperature (anodizing without thermo-chemical treatment) and a constant electrolyte temperature for the samples subjected to thermo-chemical treatment after the anodizing process. A ternary electrolyte consisting of an 18% sulphuric acid solution (33 mL/L), oxalic acid (30 g/L), and phthalic acid (76 g/L) was used for the anodizing process. During anodizing, the electrolyte was mixed mechanically at 100 rpm, the direction of rotation was changed every 10 min.

The studies were carried out according to a statistically determined poly-selective experiment plan. Hartley's plan was applied for three input factors based on a hypercube, for which coefficient α = 1. Table 1 presents the list of factors on a natural and standard scale for coatings produced in the anodizing process without thermo-chemical treatment.

Table 1. List of factors on natural and standard scale for coatings produced in anodizing process without thermo-chemical treatment.

Sample	Controlled Factors					
	On a Natural Scale			On a Standard Scale		
	Current Density j [A/dm^2]	Electrolyte Temperature T [K]	Process Time t [min]	×1	×2	×3
01A	2	293	90	−1	−1	1
01B	4	293	30	1	−1	−1
01C	2	303	30	−1	1	−1
01D	4	303	90	1	1	1
01E	2	298	60	−1	0	0
01F	4	298	60	1	0	0
01G	3	293	60	0	−1	0
01H	3	303	60	0	1	0
01I	3	298	30	0	0	−1
01J	3	298	90	0	0	1
01K	3	298	60	0	0	0

Table 2 presents the list of factors on a natural and standard scale for coatings produced in the process of anodizing and thermo-chemical treatment. The thermo-chemical treatment was carried out after thorough rinsing of the samples in distilled water. The anodizing process was carried out at the constant electrolyte temperature of 298 K. Thermo-chemical treatment was carried out in solutions of distilled water, sodium dichromate and sodium sulphate. For each compound, the treatments lasted 60 min, the temperature of the solutions was 371 K.

Table 2. List of factors on natural and standard scale for coatings produced in anodizing process and thermo-chemical treatment.

Sample	Controlled Factors					
	On a Natural Scale			On a Standard Scale		
	Current Density j [A/dm^2]	Compound for Thermo-Chemical Treatment (Density) [g/cm^3]	Process Time t [min]	×1	×2	×3
02A	2	Water (0.998)	90	−1	−1	1
02B	4	Water (0.998)	30	1	−1	−1
02C	2	Sodium dichromate (2.52)	30	−1	1	−1
02D	4	Sodium dichromate (2.52)	90	1	1	1
02E	2	Sodium sulphate (1.46)	60	−1	0	0
02F	4	Sodium sulphate (1.46)	60	1	0	0
02G	3	Water (0.998)	60	0	−1	0
02H	3	Sodium dichromate (2.52)	60	0	1	0
02I	3	Sodium sulphate (1.46)	30	0	0	−1
02J	3	Sodium sulphate (1.46)	90	0	0	1
02K	3	Sodium sulphate (1.46)	60	0	0	0

2.2. Research Methodology

The thickness of the oxide coatings was measured by the contact method using a Dualscope MP40 instrument (Helmut Fischer GmbH + Co.KG, Sindelfingen, Germany), which uses the eddy current method for measurement. Ten measurements were taken along the entire length of the samples with three repetitions and the mean values were calculated along with deviations.

Using a Hitachi S-4700 scanning microscope (Hitachi, Tokyo, Japan), a micrograph of the morphology of the Al_2O_3 coatings was made. A magnification of 50,000× was used to observe the surface. Before further studies were carried out, the surfaces were sprayed with carbon to remove electrons struck during interaction of the microscope beam. Analysis of the chemical composition of the oxide coatings was carried out using a Noran Vantage EDS (Energy Dispersive X-ray Spectroscopy) system attached to the Hitachi S-4700 scanning microscope.

Contact angle measurements were performed using four liquids, two polar (water and glycerine) and two non-polar (α-bromonaphthalene and diiodomethane). Ten drops of each liquid were applied to each of the investigated coatings using an 0.5 µL micropipette over the entire length of the sample. Photographs of each drop were taken with a camera, which were then exported to a computer. The software used allowed the marking of three extreme points on the drop photograph, which enabled automatic measurement of the contact angle. The smallest and largest angles were rejected, and the other eight angle values were used to calculate the average contact angle of a particular coating. Surface free energy was calculated by the Owens–Wendt method using one polar (water) and one non-polar liquid (α-bromonaphthalene). Calculations using the Owens–Wendt method were carried out based on the equation:

$$\gamma_s = \gamma_s^d + \gamma_s^p \tag{1}$$

where γ_s is the surface free energy of the solid, γ_s^d is the surface free energy (SFE) dispersion component of the examined materials, and γ_s^p is the SFE polar component of the investigated materials.

3. Results and Discussion

Table 3 presents the average results of thickness measurements of the oxide coatings produced in the anodizing process (without thermo-chemical treatment).

Table 3. List of Al_2O_3 coating thicknesses produced in anodizing process (without thermo-chemical treatment).

Sample	Oxide Layers Thickness d [µm]	Deviation [µm]
01A	51.3	0.94
01B	33.1	0.90
01C	16.9	0.22
01D	94.8	0.66
01E	32.6	0.45
01F	67.6	0.42
01G	53.1	1.12
01H	51.2	0.91
01I	25.2	0.33
01J	75.9	0.47
01K	51.4	0.96

The measurements showed significant changes in the thickness of the oxide coatings resulting from the use of different anodizing parameters. The thickness of the coatings depends on the process time, current density and electrolyte temperature. An increase in anodizing time at a constant current density and electrolyte temperature causes a significant increase in the coating thickness (samples 01I, 01J, 01K). As the current density increased, while the electrolyte time and temperature remained constant, a significant increase in coating thickness was also observed (samples 01E, 01F, 01K). In both cases, the increase in coating thickness is due to the increasing value of the electric charge. An increase in electrolyte temperature, while maintaining a constant anodizing time and current density, contributes to a decrease in coating thickness, which is caused by an increase in Al_2O_3 secondary solubility along with an increase in electrolyte temperature (samples 01G, 01H, 01K).

Table 4 presents the average results of thickness measurements of oxide coatings produced in the anodizing process and thermo-chemical treatment.

Table 4. Summary of Al_2O_3 coating thickness produced in anodizing process and thermo-chemical treatment.

Sample	Oxide Layers Thickness d [µm]	Deviation [µm]
02A	51.1	0.58
02B	33.5	0.81
02C	16.4	0.31
02D	98.6	5.79
02E	34.0	0.53
02F	68.7	0.59
02G	48.9	0.57
02H	51.9	0.45
02I	27.5	0.47
02J	77.1	1.31
02K	52.1	0.19

Comparing the measurements of the thickness of the oxide coatings before and after the thermo-chemical treatment produced under the same anodizing conditions, it can be stated that the treatment causes a slight increase in the thickness of the oxide coating (comparison of samples 01I, 01J, 01K with samples 02 I, 02J, 0.2 K). Confirmation of this can also be seen by comparing samples 01E, 01F, 01K with samples 02E, 02F, 02K. The increase in the layer thickness after thermo-chemical treatment is insignificant and varies within 1.5 µm. As in the case of unmodified coatings, after the thermo-chemical treatment, the dependence of the increase in Al_2O_3 thickness on the increase in current density and the time of the anodizing process can also be seen. Table 5 shows the contact angle values measured using distilled water and α-bromonaphthalene. Table 6 shows the contact angle values measured using glycerine and diiodomethane. The measurements were made on coatings produced during anodizing without thermo-chemical treatment. Depending on the anodizing parameters, different contact angle values were obtained. The highest contact angle for polar liquids (distilled water, glycerine) was obtained for the 01D sample produced at the current density of 4 A/dm^2 for 90 min at the electrolyte temperature of 303 K. The lowest value was obtained for the 01J sample produced at the current density of 3 A/dm^2 for 90 min at the electrolyte temperature of 298 K. Considering the values of the contact angles of the coatings measured using glycerine, sample 01D is a hydrophobic surface with a contact angle of 98.06° ± 2.62°.

Table 7 contains the contact angle values measured using distilled water and α-bromonaphthalene. Table 8 shows the contact angle values measured using glycerine and diiodomethane. The Al_2O_3 coatings underwent thermo-chemical treatment after previous anodizing. The highest value of contact angle measured with distilled water was obtained for sample 02E produced at the current density of 2 A/dm^2 for 60 min, then subjected to thermo-chemical treatment in sodium sulphate. The smallest contact angle value was obtained for the 02D sample produced at the current density of 4 A/dm^2 for 90 min, subjected to thermo-chemical treatment in sodium dichromate. The produced coating was highly hydrophilic—the measured angle was only 8.62° ± 2.02°. Considering the second polar liquid (glycerine), the smallest contact angle value was also measured for the 02D sample; however, the largest value was as much as 109.82° ± 4.79°—the highly hydrophobic surface was measured on the 02K sample produced at the current density of 3 A/dm^2 for 60 min and then subjected to thermo-chemical treatment in sodium sulphate. Sample 02E also showed hydrophobic properties—its contact angle was 92.78 ± 6.62°. Thermo-chemical treatment of the Al_2O_3 coatings contributed to a significant change in the wettability of the measured surfaces. In the case of the sample anodized at the current density of 3 A/dm^2 for 60 min, the thermo-chemical treatment in sodium sulphate contributed to an increase in the contact angle from 73.86° to 81.14° for water and from 71.95° to 109.82° for glycerine, creating a highly hydrophobic surface. Furthermore, the thermo-chemical treatment of samples 01E, 01F, 01K in sodium sulphate contributed to a significant reduction in wettability (a higher contact angle) for glycerine. A reduction in the electrolyte temperature during anodizing from 303 to 298 K and the thermo-chemical treatment process in sodium dichromate of samples 01C, 01D, 01H contributed to a very large increase in the wettability of the coatings, creating surfaces with strong hydrophilic properties, both using water and glycerine. An increase in the electrolyte temperature during anodizing from 293 to 298 K for samples 01A, 01B, 01G and the thermo-chemical treatment process using water resulted in a significant reduction in the contact angle.

Table 9 presents the surface free energy values calculated by the Owens–Wendt method. Wetting angles measured using distilled water and α-bromonaphthalene using the sessile drop technique were used for the calculations. The angles were measured on Al$_2$O$_3$ coating without thermo-chemical treatment.

Table 5. Wetting angles of oxide coatings produced in the anodizing process (without thermo-chemical treatment) for distilled water and α-bromonaphthalene.

Sample	Contact Angle (Distilled Water) [°]	Deviation [°]	Contact Angle (α-Bromonaphthalene) [°]	Deviation [°]
01A	74.60	2.22	22.70	3.28
01B	81.61	7.07	33.54	4.26
01C	72.55	6.35	28.82	3.22
01D	85.84	3.36	33.76	7.86
01E	76.80	2.26	27.26	1.34
01F	80.76	5.78	36.25	2.23
01G	84.06	5.71	27.51	3.46
01H	79.23	4.80	27.09	1.83
01I	80.52	5.94	26.69	1.34
01J	69.68	3.57	31.88	4.93
01K	73.86	2.83	24.97	4.12

Table 6. Wetting angles of oxide coatings produced in the anodizing process (without thermo-chemical treatment) for glycerine and diiodomethane.

Sample	Contact Angle (Glycerin) [°]	Deviation [°]	Contact Angle (Diiodomethane) [°]	Deviation [°]
01A	72.10	3.54	44.95	6.56
01B	80.67	2.09	50.44	3.09
01C	83.09	6.06	47.39	1.53
01D	98.06	2.62	64.09	5.07
01E	77.38	1.23	47.87	3.52
01F	85.98	5.29	47.96	4.34
01G	79.55	3.17	49.66	6.29
01H	76.68	4.08	49.16	3.56
01I	79.75	3.45	48.48	1.95
01J	71.19	2.81	52.65	2.51
01K	71.95	4.70	51.68	3.23

Table 7. Wetting angles of oxide coatings produced in anodizing process and thermo-chemical treatment for distilled water and α-bromonaphthalene.

Sample	Contact Angle (Distilled Water) [°]	Deviation [°]	Contact Angle (α-Bromonaphthalene) [°]	Deviation [°]
02A	56.81	2.48	13.93	2.64
02B	60.44	1.97	17.67	2.62
02C	27.44	4.13	15.89	1.63
02D	8.62	2.02	9.54	2.74
02E	89.61	6.34	44.83	5.26
02F	63.18	3.55	22.15	3.21
02G	71.29	11.02	18.89	3.56
02H	22.55	1.95	14.44	2.81
02I	61.74	4.15	38.33	3.88
02J	52.77	6.94	22.63	2.77
02K	81.14	6.16	37.86	3.63

Table 8. Wetting angles of oxide coatings produced in the anodizing process and thermo-chemical treatment for glycerine and diiodomethane.

Sample	Contact Angle (Glycerin) [°]	Deviation [°]	Contact Angle (Diiodomethane) [°]	Deviation [°]
02A	73.84	5.01	30.88	5.52
02B	84.47	4.97	34.57	1.90
02C	54.11	7.56	32.05	3.35
02D	27.67	2.53	17.88	2.47
02E	92.78	6.62	59.71	10.68
02F	88.74	7.26	36.89	3.22
02G	85.40	6.81	37.24	5.47
02H	41.19	5.01	23.33	3.86
02I	88.25	3.05	52.99	6.19
02J	64.60	4.74	41.79	3.43
02K	109.82	4.79	57.25	5.38

Table 9. Surface free energy values of Al_2O_3 coatings without thermo-chemical treatment.

Sample	SFE Owens-Wendt [mJ/m^2]
01A	43.15
01B	41.35
01C	46.12
01D	40.06
01E	44.85
01F	40.73
01G	37.66
01H	39.56
01I	43.75
01J	46.57
01K	40.84

The highest value of surface free energy was determined for the 01J sample, i.e., for the coating characterized by the smallest contact angle for water, while the lowest SFE value was determined for the surface characterized by one of the highest contact angles for water (84.06°)—sample 01G. After the calculations, it can be stated that on oxide coatings without thermo-chemical treatment, the contact angles show inverse proportionality to the value of surface free energy. Three-dimensional graphs were made to visualize the effect of the anodizing parameters on SFE (Figure 1).

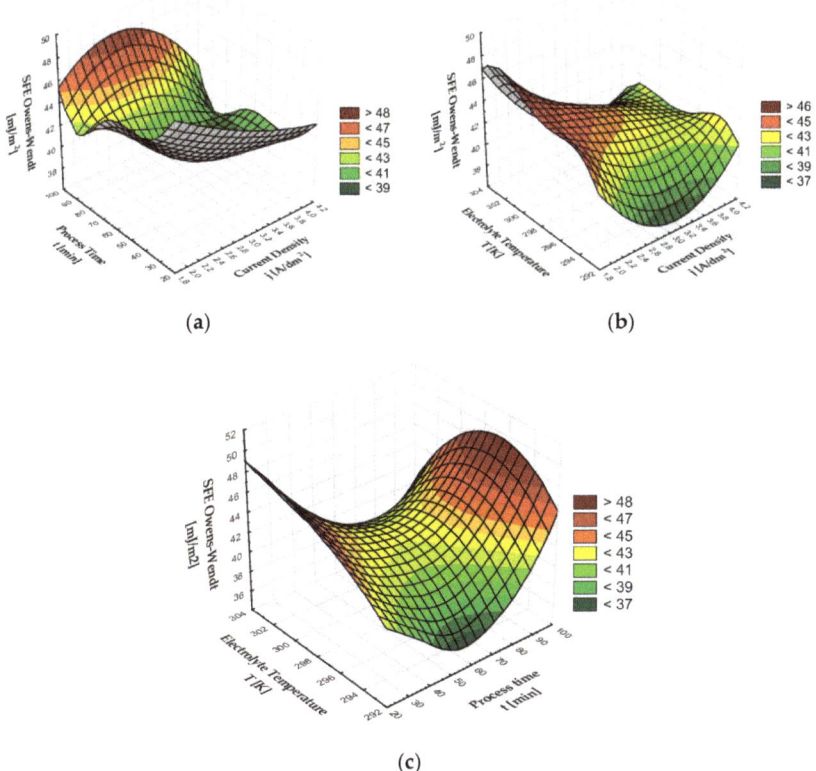

Figure 1. Surface free energy (SFE) dependence on: (**a**) process time and current density, (**b**) electrolyte temperature and current density, (**c**) electrolyte temperature and process time.

By analyzing the impact of the process time and current density on SFE, it can be seen that the highest values of surface free energy of over 44 [mJ/m^2] were exhibited by a coating produced within 100 min and at the current density of 3 A/dm^2. Values over 40 [mJ/m^2] were also observed for coatings produced within 30 min for all the current density values. The graph showing the effect of electrolyte temperature and current density on SFE shows a clear increase in the surface free energy value in the middle of the electrolyte temperature axis for a current density of about 2 A/dm^2. The lowest values < 39 [mJ/m^2] were attained for the current density of 3 A/dm^2 and electrolyte temperature around 293 K. The next graph presents the dependence of SFE on the electrolyte temperature and process time. The highest values > 44 [mJ/m^2] were found at the beginning and end of the process time axis at values in the middle of the electrolyte temperature axis (298 K).

Table 10 shows the SFE calculated for coatings after thermo-chemical treatment. The calculations were carried out using the Owen–Wendt method based on the contact angles of distilled water and α-bromonaphthalene.

Table 10. Values of surface free energy of Al_2O_3 coatings after thermo-chemical treatment.

Sample	SFE Owens-Wendt [mJ/m^2]
02A	56.60
02B	54.19
02C	71.73
02D	77.94
02E	34.98
02F	52.01
02G	48.91
02H	73.85
02I	48.92
02J	57.49
02K	40.06

Sample 02D showed the highest surface free energy; this is a coating which, after anodizing, was subjected to thermo-chemical treatment in sodium dichromate. The coating also has the lowest contact angle for water (8.62°) and for glycerine (27.67°). The lowest SFE value was calculated for sample 02E, which has the highest contact angle for water (89.61°) and one of the highest for glycerine (92.78°). As with the Al_2O_3 coatings without modification, the contact angles of the coatings after thermo-chemical treatment show an inverse proportionality to the value of surface free energy. Figure 2 presents 3D graphs of the dependence of the anodizing parameters and compounds used for thermo-chemical treatment on the SFE value.

The surface free energy of coatings after thermo-chemical treatment varies depending on the process time and current density. The highest SFE values > 75 mJ/m^2 were found at the intersection of the end of the process time axis (about 90 min) and the end of the current density axis (about 4 A/dm^2). A clear increase can also be seen at the intersection of a current density of about 2 A/dm^2 and a process time of about 30 min. By analyzing the impact of the current density and the compound used for thermochemical treatment, it can be stated that the greatest impact on the increase in the SFE value is the use of sodium dichromate in the coating modification process. When using sodium dichromate, a clear increase in surface free energy is visible in the entire current density range, with a peak value > 76 mJ/m^2 for a current density of about 4 A/dm^2. A clear decrease in the SFE value can be seen for sodium sulphate at the current density of 2 A/dm^2. The graph showing the dependence of SFE on the time of the process and the compound used for thermochemical treatment shows a clear maximum SFE value > 76 mJ/m^2 for sodium dichromate, decreasing with the time of the process. Surface free energy values > 56 mJ/m^2 are also visible in the case of anodized coatings that underwent thermo-chemical treatment in water for about 90 min. A clear decrease in the SFE value < 44 mJ/m^2 is visible in the middle of the graph for anodized coatings modified in the sodium sulphate solution for about 60 min.

Figure 3 presents surface morphology micrographs taken with a scanning microscope of selected oxide coatings with the largest differences in the contact angle and SFE. A 50,000× magnification was used for all the morphology micrographs.

The micrographs show surface porosity, which is characteristic of Al_2O_3 coatings. There are significant differences in the size of the nanopores, which depend on both the anodizing time and the current density used in the anodizing process.

Figure 4 presents micrographs of the surface morphology of layers anodized and thermo-chemically treated in sodium dichromate and sodium sulphate.

The thermo-chemical treatment of the 02D sample carried out in sodium dichromate contributed to complete sealing of the porous Al_2O_3 coating and the formation of a high sodium content sublayer. On the surfaces of the 02E and 02K samples, which after the anodizing process were subjected to thermo-chemical treatment in sodium sulphate, complete coverage of the oxide coating surface with a kind of crosslinked structure is noticeable.

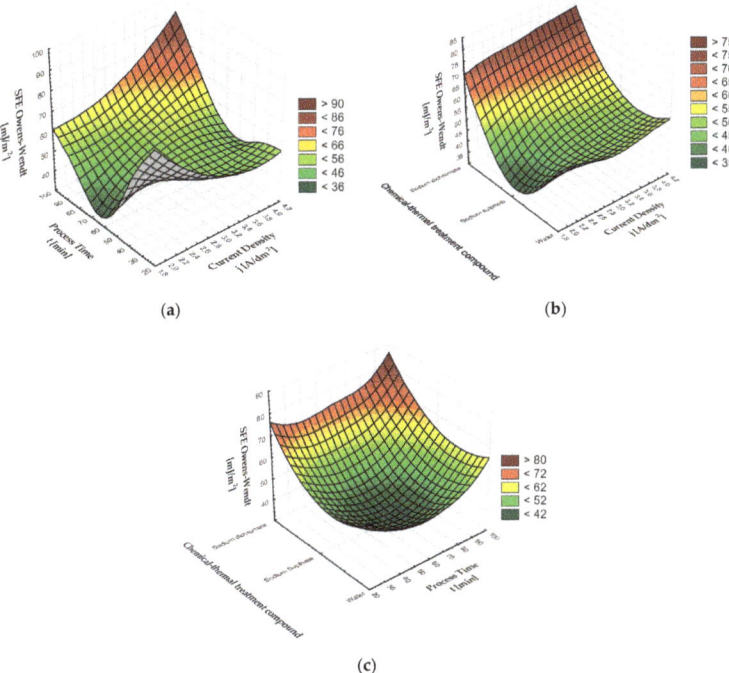

Figure 2. SFE dependence on: (**a**) process time and current density, (**b**) chemical compounds used in coating modification and current density, (**c**) chemical compounds used in coating modification and process time.

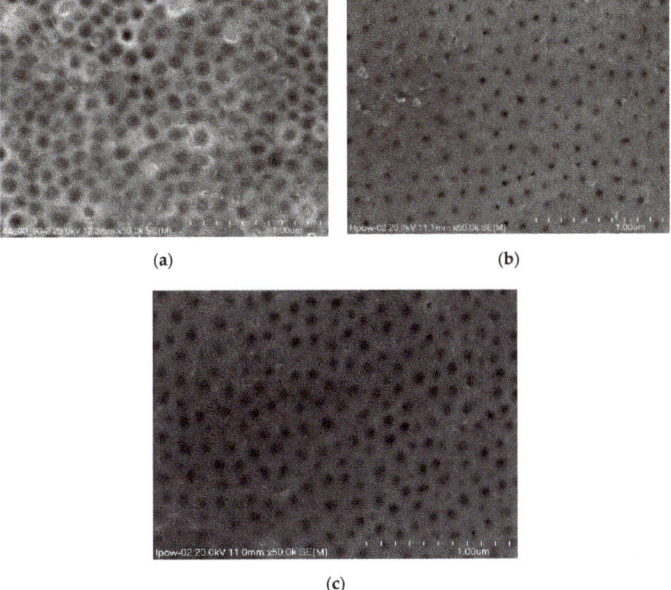

Figure 3. Surface morphology micrographs of oxide coatings without thermo-chemical treatment: (**a**) Sample 01D, (**b**) Sample 01H, (**c**) Sample 01J.

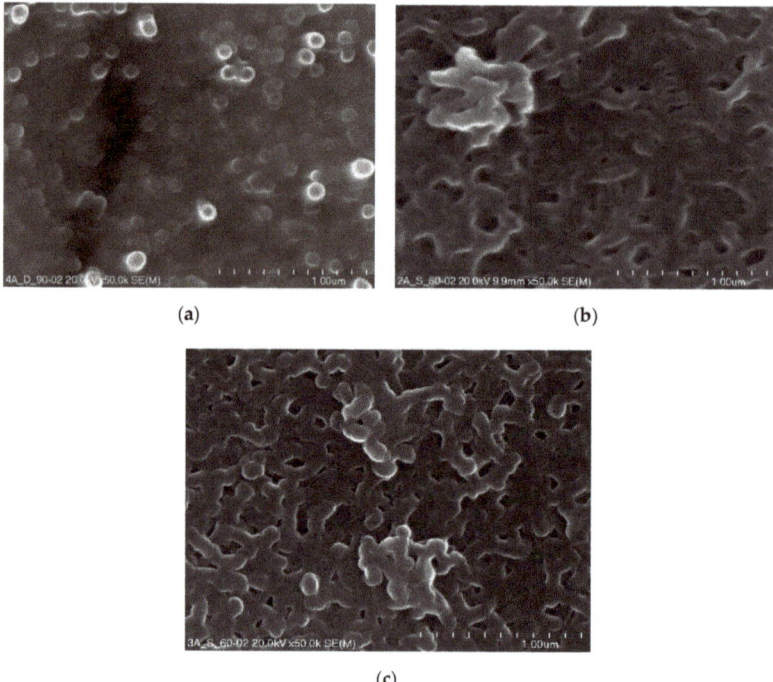

(a) (b) (c)

Figure 4. Surface morphology micrographs of oxide layers after thermo-chemical treatment: (**a**) Sample 02D, (**b**) Sample 02E, (**c**) Sample 02K.

Table 11 summarizes the chemical composition of the oxide coatings produced in the DC anodizing process without the thermo-chemical process.

Table 11. Chemical composition analysis of Al_2O_3 coatings without thermo-chemical process.

Sample	Atomic Aluminum Content [%]	Error of Aluminum Content [%]	Atomic Oxygen Content [%]	Error of Oxygen Content [%]
01D	54.52	±0.30	43.94	±0.68
01H	56.23	±0.21	44.25	±0.54
01J	55.90	±0.23	43.49	±0.51

Samples with significant differences in the values of surface free energy and contact angle were selected for analysis. The atomic content of both aluminum and oxygen in the selected coatings are very similar. An increase in aluminum content was observed for the samples with a smaller Al_2O_3 coating thickness. The increased aluminum content is due to the shorter distance between the coating surface and the substrate. The values presented in the table are very similar to stoichiometric calculations of the alumina chemical composition.

Figure 5 presents SEM micrographs for samples 02D, 02E, 02K together with the marked field used for chemical composition analysis. Figure 6 shows the EDS spectrum for the 02D, 02E, 02K samples with chemical composition analysis.

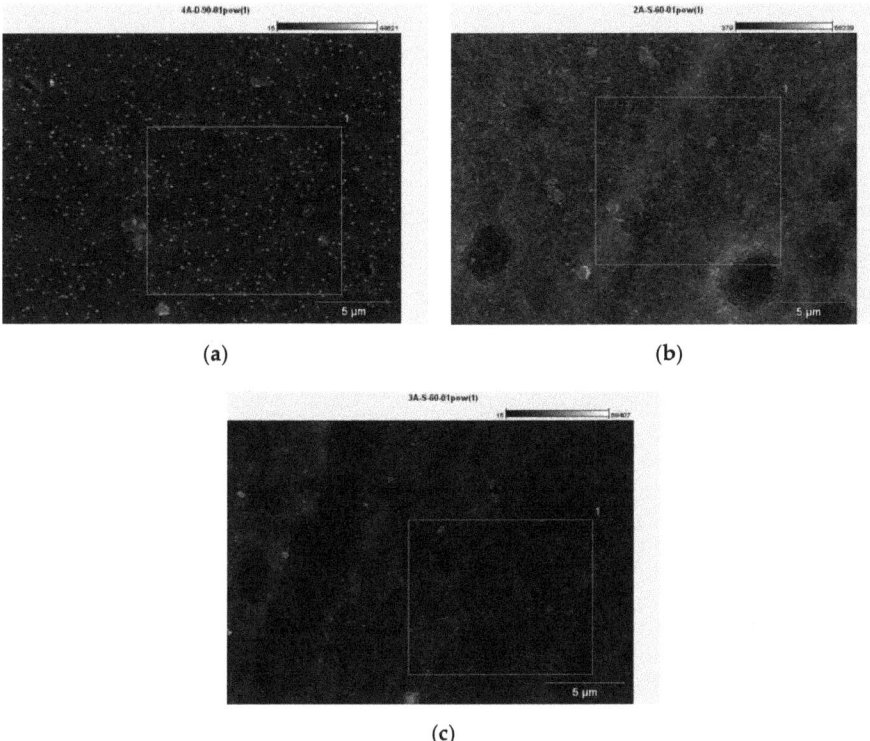

Figure 5. SEM micrographs with chemical analysis field marked: (**a**) Sample 02D, (**b**) Sample 02E, (**c**) Sample 02K.

The coating composition of sample 02D is characterized by a high content of oxygen, at the same time reducing the content of aluminum. Thermo-chemical treatment in sodium dichromate contributed to the building of over 18% sodium compounds on the surface. On the surfaces of the 02E and 02K samples modified in sodium sulphate, a reduction in oxygen content for about 2% sulphur was shown.

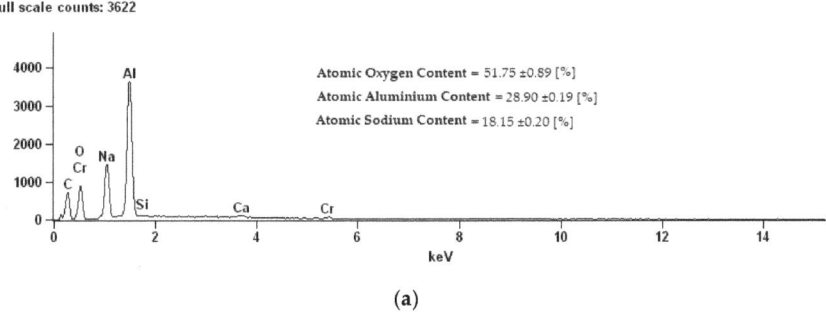

(**a**)

Figure 6. *Cont.*

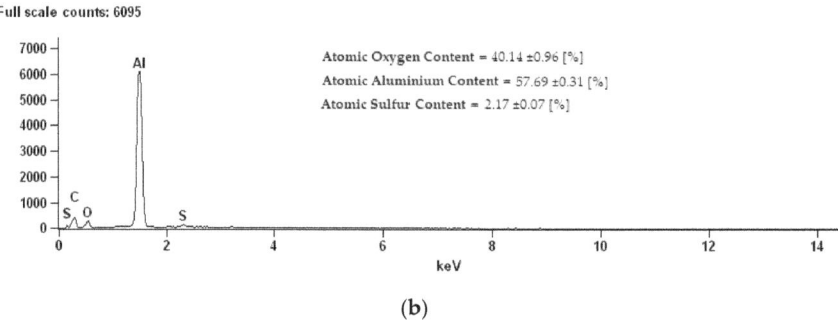

(b)

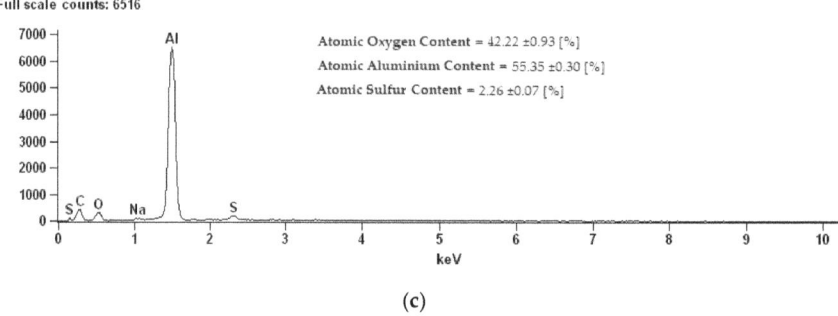

(c)

Figure 6. EDS (Energy Dispersive X-ray Spectroscopy) spectrum along with chemical analysis for samples: (**a**) Sample 02D, (**b**) Sample 02E, (**c**) Sample 02K.

4. Conclusions

Based on the conducted studies, it can be concluded that both the anodizing parameters and thermo-chemical treatment have a significant impact on the wettability of Al_2O_3 coating surfaces, and thus their energy state. By controlling the anodizing parameters (current density, electrolyte temperature, process time), a hydrophobic surface with a contact angle of 98.06° ± 2.62° was obtained for the 01D sample using glycerine as the measuring liquid. The 01D sample oxide coating also achieved the highest contact angle with distilled water. The coating was produced at the current density of 4 A/dm^2 for 90 min at the electrolyte temperature of 303 K. The lowest contact angle value was obtained by the coating produced at the current density of 3 A/dm^2 for 90 min at the electrolyte temperature of 298 K—sample 01J.

The second part of the research focused on the modification of the Al_2O_3 coatings by chemical compounds through thermo-chemical treatment. A very clear effect of thermo-chemical treatment on the values of contact angles was noticed. The highest contact angle value measured using distilled water was obtained by sample 02E modified in sodium sulphate. The smallest contact angle value was obtained by a sample modified in sodium dichromate. The coating modified in sodium dichromate showed strongly hydrophilic values; the measured angle was only 8.62° ± 2.02°. For the 02K sample modified in sodium sulphate, the contact angle measurements with glycerine revealed an angle of 109.82° ± 4.79° (a strongly hydrophobic surface). Moreover, sample 02E exhibited hydrophobic properties; the contact angle for this coating was 92.78° ± 6.62°. The thermo-chemical treatment in sodium sulphate of the 02K sample increased the contact angle from 73.86° to 81.14° for water and from 71.95° to 109.82° for glycerine, creating a highly hydrophobic surface. The modification also contributed to a significant reduction in wettability for glycerine. The reduction in electrolyte temperature in the anodizing process of the coatings modified in sodium dichromate contributed to a very large increase in the wettability of the coatings, creating surfaces with strong hydrophilic properties, both when measuring with distilled water and glycerine. The increase in electrolyte temperature during anodizing from 293 to 298 K and the thermo-chemical treatment process using water resulted in a significant reduction in the contact angle measured for distilled water. The studies have also shown that the contact angle is inversely related to surface free energy.

Author Contributions: M.B. performed the analysis of the results; M.N. carried out the tests and wrote the manuscript; W.S. contributed to the concept and modified the manuscript. All authors have read and agreed to the published version of the manuscript.

Funding: This research received no external funding.

Conflicts of Interest: The authors declare no conflict of interest.

References

1. Hatch, J.E. *Aluminum: Properties and Physical Metallurgy*; ASM International: Cleveland, OH, USA, 1984.
2. Khoei, A.R.; Maters, I.; Gethin, D.T. Design optimisation of aluminium recycling processes using Taguchi technique. *J. Mater. Process. Technol.* **2002**, *127*, 96–106. [CrossRef]
3. Vargel, C. *Corrosion of Aluminium*; Elsevier Science: Amsterdam, The Netherlands, 2004.
4. Davis, J.R. *Corrosion of Aluminium and Aluminium Alloys*; ASM International: Novelty, OH, USA, 1999.
5. Hirsch, J.; Skrotzki, B.; Gottstein, G. *Aluminum Alloys: Their Physical and Mechanical Properties*; Wiley-VCH Cerlag Gmbh & Co. KGaA: Weinheim, Germany, 2008.
6. Mondolfo, L.F. *Aluminum Alloys: Structure and Properties*; Elsevier Science: London, UK, 2013.
7. Starke, E.A.; Staley, J.T. Application of modern aluminium alloys to aircraft. *Prog. Aerosp. Sci.* **1996**, *32*, 131–172. [CrossRef]
8. Miller, W.S.; Zhuang, L.; Bottema, J.; Witterbrood, A.J.; De Smet, P.; Haszler, A.; Vieregge, A. Recent development in aluminium alloys for the automotive industry. *Mater. Sci. Eng. A-Struct.* **2000**, *280*, 37–49. [CrossRef]
9. Calcraft, R.C.; Wahab, R.A.; Bottema, J.; Viano, D.M.; Schumann, G.O.; Phillips, R.H.; Ahmed, N.U. The development of the welding procedures and fatigue of butt-welded structures of aluminium-AA5383. *J. Mater. Process. Technol.* **1999**, *92–93*, 60–65. [CrossRef]
10. Poznak, A.; Freiberg, D.; Sanders, P. Chapter 10—Automotive Wrought Aluminium Alloys. In *Fundamentals of Aluminium Metallurgy*; Lumley, R.N., Ed.; Woodhead Publishing: Sawston, UK, 2018; pp. 333–386.
11. Kök, M.; Özdin, K. Wear resistance of aluminium alloy and its composites reinforced by Al_2O_3 particles. *J. Mater. Process. Technol.* **2007**, *183*, 301–309. [CrossRef]
12. Stepniowski, W.J.; Moneta, M.; Karczewski, K.; Michalska-Domanska, M.; Czujko, T.; Mol, J.M.; Buijnsters, J.G. Fabrication of copper nanowires via electrodeposition in anodic aluminum oxide templates formed by combined hard anodizing and electrochemical barrier layer thinning. *J. Electroanal. Chem.* **2018**, *809*, 59–66. [CrossRef]
13. Niedźwiedź, M.; Skoneczny, W.; Bara, M. The influence of anodic alumina coating nanostructure produced on EN AW-5251 alloy on type of tribological wear process. *Coatings* **2020**, *10*, 105. [CrossRef]
14. Bara, M.; Kmita, T.; Korzekwa, J. Microstructure and properties of composite coatings obtained on aluminium alloys. *Arch. Metall. Mater.* **2016**, *61*, 1107–1112. [CrossRef]
15. Kok, M. Production and mechanical properties of Al_2O_3 particle-reinforced 2024 aluminium alloy composites. *J. Mater. Process. Technol.* **2005**, *161*, 381–387. [CrossRef]
16. Kwolek, P.; Drapała, D.; Krupa, K.; Obłój, A.; Tokarski, T.; Sieniawski, J. Mechanical properties of a pulsed anodised 5005 aluminium alloy. *Surf. Coat. Technol.* **2020**, *383*, 1–8. [CrossRef]
17. Zhao, N.-Q.; Jiang, X.-X.; Shi, C.-S.; Li, J.-J.; Zhao, Z.-G.; Du, X.-W. Effects of anodizingconditions on anodic alumina structure. *J. Mater. Sci.* **2007**, *42*, 3878–3882. [CrossRef]
18. Sulka, G.D. *Highly Ordered Anodic Porous Alumina Formation by Self-Organized Anodizing*; Wiley-VCH Verlag GmbH & Co., KGaA: Weinheim, Germany, 2008.
19. Liu, P.; Singh, V.P.; Rajaputra, S. Barrier layer nonuniformity effects in anodized aluminum oxide nanopores on ITO substrates. *Nanotechnology* **2010**, *21*, 115303. [CrossRef] [PubMed]
20. Zhang, L.; Cho, H.S.; Li, F.; Metzger, R.M.; Doyle, W.D. Cellular growth of highly ordered porous anodic films on aluminium. *J. Mater. Sci. Lett.* **1998**, *17*, 291–294. [CrossRef]

21. Aerts, T.; Dimogerontakis, T.; De Graeve, I.; Fransaer, J.; Terryn, H. Influence of the anodizing temperature on the porosity and the mechanical properties of the porous anodic oxide film. *Surf. Coat. Technol.* **2007**, *201*, 7310–7317. [CrossRef]
22. Hao, L.; Cheng, B.R. Sealing processes of anodic coatings—Past, present, and future. *Met. Finish.* **2000**, *98*, 8–18. [CrossRef]
23. Kyungtae, K.; Moonjung, K.; Sung, M.C. Pulsed electrodeposition of palladium nanowire arrays using AAO template. *Mater. Chem. Phys.* **2006**, *98*, 278–282.
24. Bartolome, M.J.; Lopez, V.; Escudero, E.; Caruana, G.; Gonzales, J.A. Changes in the specific surface area of porous aluminium oxide films during sealing. *Surf. Coat. Technol.* **2006**, *200*, 4530–4537. [CrossRef]
25. Hoar, T.P.; Wood, G.C. The sealing of porous anodic oxide films on aluminium. *Electrochim. Acta* **1962**, *7*, 333–353. [CrossRef]
26. Whelan, M.; Cassidy, J.; Duffy, B. Sol–gel sealing characteristics for corrosion resistance of anodised aluminium. *Surf. Coat. Technol.* **2013**, *235*, 86–96. [CrossRef]
27. Bara, M.; Niedźwiedź, M.; Skoneczny, W. Influence of anodizing parameters on surface morphology and surface-free energy of Al_2O_3 layers produced on EN AW-5251 alloy. *Materials* **2019**, *12*, 695. [CrossRef]
28. Zhang, X.; Shi, F.; Niu, J.; Jiang, Y.G.; Wang, Z.Q. Superhydrophobic surfaces: From structural control to functional application. *J. Mater. Chem.* **2008**, *18*, 621–633. [CrossRef]
29. Wang, Q.; Zhang, B.W.; Qu, M.N.; Zhang, J.Y.; He, D.Y. Fabrication of superhydrophobic surfaces on engineering material surfaces with stearic acid. *Appl. Surf. Sci.* **2008**, *254*, 2009–2012. [CrossRef]
30. Xu, X.; Zhu, Y.; Zhang, L.; Sun, J.; Huang, J.; Huang, J.; Chen, J.; Cao, Y. Hydrophilic poly(triphenylamines) with phosphonate groups on the side chains: Synthesis and photovoltaic applications. *J. Mater. Chem.* **2012**, *22*, 4329–4336. [CrossRef]
31. Zhang, Z.; Wu, Q.; Song, K.; Ren, S.; Lei, T.; Zhang, Q. Using cellulose nanocrystals as a sustainable additive to enhance hydrophilicity, mechanical and thermal properties of poly(vinylidene fluoride)/poly(methyl methacrylate) blend. *ACS Sustain. Chem. Eng.* **2015**, *3*, 574–582. [CrossRef]
32. Li, X.; Liu, K.L.; Wang, M.; Wong, S.Y.; Tjiu, W.C.; He, C.B.; Goh, S.H.; Li, J. Improving hydrophilicity, mechanical properties and biocompatibility of poly[(R)-3-hydroxybutyrate-co-(R)-3-hydroxyvalerate] through blending with poly[(R)-3-hydroxybutyrate]-alt-poly(ethylene oxide). *Acta Biomater.* **2009**, *5*, 2002–2012. [CrossRef]
33. Ohtsu, N.; Hirano, Y. Growth of oxide layers on NiTi alloy surfaces through anodization innitric acid electrolyte. *Surf. Coat. Technol.* **2017**, *325*, 75–80. [CrossRef]
34. Mokhtari, S.; Karimzadeh, F.; Abbasi, M.H.; Raeissi, K. Development of super-hydrophobic surface on Al 6061 by anodizing and the evaluation of its corrosion behavior. *Surf. Coat. Technol.* **2017**, *324*, 99–105. [CrossRef]
35. Buijnters, J.G.; Zhong, R.; Tsyntsaru, N.; Celis, J.-P. Surface wettability of macroporous anodized aluminum oxide. *ACS Appl. Mater. Interfaces* **2013**, *5*, 3224–3233. [CrossRef]
36. Kim, D.; Hwang, W.; Park, H.C.; Lee, K.-H. Superhydrophobic nanostructures based on porous alumina. *Curr. Appl. Phys.* **2007**, *8*, 770–773. [CrossRef]
37. Tasaltin, N.; Sanli, D.; Jonáš, A.; Kiraz, A.; Erkey, C. Preparation and characterization of superhydrophobic surfaces based on hexamethyldisilazane-modified nanoporous alumina. *Nanoscale Res. Lett.* **2011**, *6*, 487. [CrossRef]
38. Wang, H.; Dai, D.; Wu, C. Fabrication of superhydrophobic surfaces on aluminum. *Appl. Surf. Sci.* **2008**, *254*, 5599–5601. [CrossRef]

© 2020 by the authors. Licensee MDPI, Basel, Switzerland. This article is an open access article distributed under the terms and conditions of the Creative Commons Attribution (CC BY) license (http://creativecommons.org/licenses/by/4.0/).

Article

The Influence of Anodic Alumina Coating Nanostructure Produced on EN AW-5251 Alloy on Type of Tribological Wear Process

Mateusz Niedźwiedź, Władysław Skoneczny * and Marek Bara

Institute of Materials Engineering, Faculty of Science and Technology, University of Silesia in Katowice, 40-007 Katowice, Poland; mateusz.niedzwiedz@us.edu.pl (M.N.); marek.bara@us.edu.pl (M.B.)
* Correspondence: wladyslaw.skoneczny@us.edu.pl; Tel.: +48-32-368563

Received: 11 December 2019; Accepted: 21 January 2020; Published: 24 January 2020

Abstract: The article presents the influence of the anodic alumina coating nanostructure produced on aluminum alloy EN AW-5251 on the type of tribological wear process of the coating. Oxide coatings were produced electrochemically in a ternary electrolyte by the DC method. Analysis of the nanostructure of the coating was performed using ImageJ 1.50i software on micrographs taken with a scanning electron microscope (SEM). Scratch tests of the coatings were carried out using a Micron-Gamma microhardness tester. The scratch marks were subjected to surface geometric structure studies with a Form TalySurf 2 50i contact profiler. Based on the studies, it was found that changes in the manufacturing process conditions (current density, electrolyte temperature) affect changes in the coating thickness and changes in the anodic alumina coating nanostructure (quantity and diameter of nanofibers), which in turn has a significant impact on the type of tribological wear. An increase in the density of the anodizing current from 1 to 4 A/dm^2 causes an increase in the diameter of the nanofibers from 75.99 ± 7.7 to 124.59 ± 6.53 nm while reducing amount of fibers from 6.6 ± 0.61 to 3.8 ± 0.48 on length 1 × 10^3 nm. This affects on a change in the type of tribological wear from grooving to micro-cutting.

Keywords: aluminum oxide layers; nanostructure; tribological wear

1. Introduction

Aluminum alloys are currently widely used in industry due to the low weight of the metal in relation to high mechanical strength and good thermal conductivity [1,2]. Due to the atmosphere, aluminum is automatically covered by a passive oxide coating a few nanometers thick, isolating the metal from contact with the environment [3]. However, in acidic or alkaline environments, the spontaneous oxide coating does not protect the metal well enough and aluminum corrosion occurs fairly quickly, leading to mass loss [4]. Therefore, ensuring proper protection of the aluminum surface involves the formation of an oxide coating of an appropriate thickness [5]. One method that allows adequate protection of aluminum alloys is electrochemical oxidation of the alloy surface. This process is called aluminum anodizing, and due to its properties that improve the hardness of metal, it is used on a large scale today [6]. Anodic alumina coatings in the electrochemical process can be formed using DC [7], constant voltage [8], pulse methods [9], and using AC [10], as well as AC imposed on DC [11]. Electrochemically produced oxide coatings have an amorphous structure [12]. The anodic alumina coating adopts a porous structure with vertically aligned cylindrical pores, depending mainly on the applied current density, electrolyte temperature, and aluminum alloy [13,14]. An important feature of anodic alumina coatings is their resistance to abrasive wear. This results in a wide application of anodized aluminum in friction pairs of engineering kinematic systems [15]. For this reason, tests are often carried out to assess the impact of the anodizing parameters on the mechanical and tribological properties of oxide coatings [16]. One of the tests determining the mechanical properties of the coatings is the scratch test. Such tests allow one to not only determine the coefficient of friction value but also to determine the value of critical load causing damage to the coating [17,18]. In the first cited article, scientists investigated scratches of anodized aluminum in 10% oxalic acid at room temperature at 10–40 V. Subsequently, the crystal structure, chemical composition, surface morphology, and surface topography were examined. Studies have shown that higher anodizing potential leads to the formation of porous, thicker, and harder anodic alumina coatings. Anodized aluminum showed the formation of horizontal and vertical parallel cracks when attempting to scratch at a load of about 1 N. In the second article cited, researchers investigated the mechanical properties of AZ91 magnesium alloy coated with a double-layer Al_2O_3/Al. Mechanical properties were examined, among others, by a scratch test. It has been found that the mechanical properties of the coated alloy are significantly better, and the coating exhibits excellent adhesion to the substrate. Nanostructure studies are an important element of research related to the wear of oxide coatings and their adhesion to the substrate [19]. In the cited article, the authors started to produce Al_2O_3 layers by hard anodizing and thermo-chemical treatment. Then, they studied the microstructure, morphology, and tribological wear. Microstructure tests showed fiber sealing after physicochemical treatment, and it also increased the microhardness and polymer consumption during the tribological test. Our work focuses on the scratching of aluminum surfaces and anodic oxide coatings produced using various manufacturing parameters. In addition to the scratch resistance test, nanostructure analysis was performed to show the relationship between changes in the nanostructure and their impact on the type of tribological wear of the coatings which has not yet been studied. Scientists studied the nanostructure and friction of oxide layers in sliding running using finite elements [20]. They focused mainly on the oxidation of aluminum alloy for 60 min in an electrolyte consisting of sulfur and oxalic adipic acid at 298 K and a current density of 3 A/dm^2. The tribological properties of the oxide layer during reciprocating friction were tested using four different polymers (TG15, TGK20/5, TMP12, PEEK/BG). The highest wear and the highest friction coefficient were found for TMP12 material, while the lowest wear and friction coefficient for PEEK/BG. However, all cited articles lack reference to nanostructure for the type of tribological process, which makes our research innovative.

2. Materials and Methods

2.1. Research Material

The research material was anodic alumina coatings produced by an electrochemical method on the surface of plates made of EN AW-5251 aluminum alloy. This aluminum alloy is characterized by high mechanical strength and corrosion resistance. The chemical composition of the EN AW-5251 alloy (Table 1) allows one to easily modify its surface by anodizing.

Table 1. Chemical composition of the EN AW-5251 alloy.

Si	Fe	Cu	Mn	Mg	Cr	Ni	Zn	Ti	V	Other	Al
max 0.4	max 0.5	max 0.15	0.1–0.5	1.7–2.4	max 0.15	–	max 0.15	max 0.15	–	max 0.05	rest

Samples with an area of 0.1 dm^2 were subjected to etching using a 5% KOH solution (20 min) and a 10% HNO_3 solution (5 min). The temperature of the solutions was 296 K. The etching treatments ended with rinsing in distilled water. The sample surfaces (except for the reference sample) were modified by anodizing to form an anodic alumina coating on their surface using an electrochemical method. The surface of the samples was anodized by the DC method, using a GPR-25H30D power supply. During anodizing, a variable current density and electrolyte temperature were used to determine how the process parameters affect the type of tribological wear. A ternary electrolyte constituting an aqueous solution of 18% sulfuric acid (33 mL/L), oxalic acid (30 g/L), and phthalic acid

(76 g/L) was used for the anodizing process. Anodic hard coatings produced in electrolytes constituting of an aqueous solution sulfuric or oxalic acid require the use of low process temperatures (273–278 K). The addition of phthalic acid ensures that hard layers are obtained and enables the process to be carried out at room temperature. The anodizing process conditions are shown in Table 2.

Table 2. Anodizing process conditions.

Sample	Current Density j [A/dm^2]	Process Time t [min]	Electrolyte Temperature T [K]
A	1	20	283
B	1	20	293
C	4	20	313
D	4	20	283

After anodizing, the samples were rinsed in distilled water.

2.2. Research Methodology

The thickness of the anodic alumina coatings was measured by the contact method using a Fischer Dualscope MP40 instrument (Helmut Fischer GmbH+Co.KG, Sindelfingen, Germany) using the eddy-current method. Ten measurements were made (repeated 3 times) over the entire length of the sample, then the average values were calculated. Nanostructure tests of the oxide coatings were carried out on metallographic specimens using a Hitachi S-4700 scanning microscope (Hitachi, Tokyo, Japan) at 30,000× magnification. Anodic alumina coatings are non-conductive material, so they charge electrically when scanning through an electron beam, which prevents correct observation of the preparations. For better observation of the nanostructure, the samples were sputtered with carbon, which enables the discharge of electrons rebound during research. The micrographs were used for computer analysis of the image carried out using ImageJ 1.50i software. Using the image analysis procedures (smooth, bandpass filter, threshold), the number of fibers per nm and the diameter of the fibers were calculated. The scratch test was performed using a Micron-Gamma device (MicronSystema, Kiev, Ukraine). A Rockwell diamond indenter was used with a tip radius of 0.2 mm. A load of 4 ± 0.01 N was used during the test. The scratch marks were subjected to surface geometric structure (SGS) analysis. For this purpose, a Form TalySurf Series 2.50i profiler was used. The systematic scanning method was used on the transverse profile. The SGS parameters allowed the authors to determine the type of tribological wear.

3. Results and Discussion

Measurements of the thickness of the oxide coatings showed significant differences in the thickness depending on the anodizing conditions. Averaged values of measurements of the anodic alumina coating thickness together with deviations are presented in Table 3.

Table 3. Influence of anodizing conditions on thickness of oxide layers formed on EN AW-5251 aluminum alloy surface.

Sample	Oxide Layers Thickness d [µm]	Deviation [µm]
A	5.5	0.59
B	7.7	0.44
C	23.6	2.07
D	26.3	1.47

The current density as well as the electrolyte temperature during the anodizing process affects the thickness of the anodic alumina coatings. Considering the current dependence, one can notice a significant increase in coating thickness at a constant electrolyte temperature and anodizing time together with an increase in current density (Samples A and D). An increase in the thickness of the coating along with an increase in current density (at a constant process time) is associated with an increase in electric charge. The anodic oxidation process, like any electrochemical process, proceeds according to Faraday's law, according to which a specific amount of electricity transforms aluminum into a specific amount of its oxide. Taking into consideration the temperature dependence at low current density values (Samples A and B), as the electrolyte temperature increases, the coating thickness increases. The increase in electrolyte temperature accelerates the migration of ions, which in turn increases the growth rate of the coating. However, at higher current density values (Samples C and D), the coating thickness decreases as the electrolyte temperature increases. Depending on the type of electrolyte in which the electrolysis process is carried out, a certain amount of produced Al_2O_3 is always dissolved by the electrolyte. In the thicker coatings, due to the resistance of the electrolyte column lying in the pores located between the oxide fibers, Joule's heat is released. The reduction in the Al_2O_3 layer thickness as the electrolyte temperature rises, at higher current densities, can therefore be attributed to the increasing secondary solubility of aluminum oxide associated with the higher electrolyte temperature.

Micrographs of the oxide coating nanostructures taken on metallographic specimens at 30,000× magnification (Figure 1) showed structure oriented along the direction of growth of the coating under the influence of an electric field.

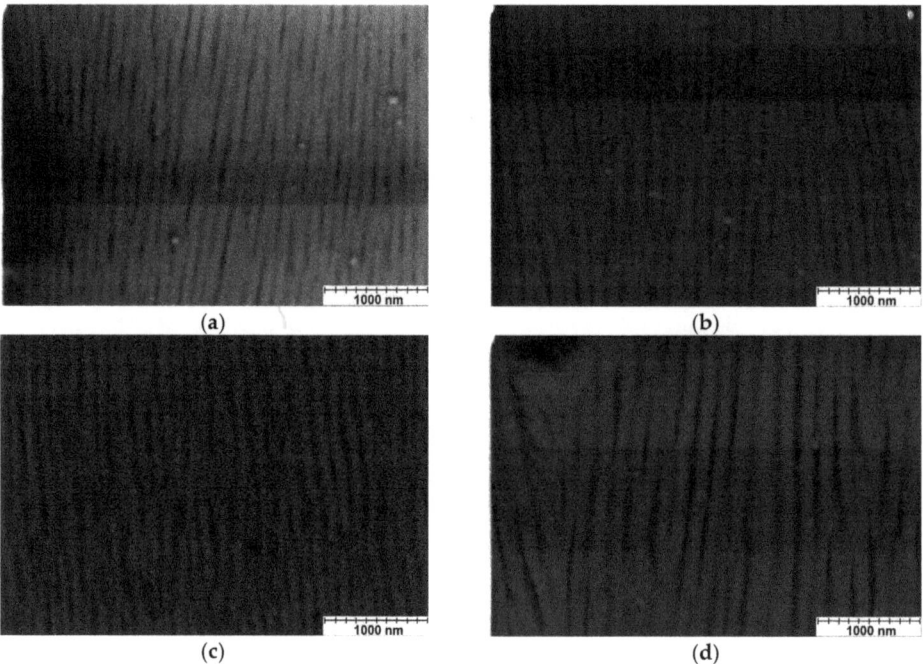

Figure 1. SEM micrograph (cross-section) of nanostructure of the coatings: (**a**) Sample A, (**b**) Sample B, (**c**) Sample C, (**d**) Sample D (sample designations as in Table 2).

In order to assess the impact of the coating production conditions on their nanostructure, the thickness and number of coating fibers were measured. For this purpose, each of the micrographs was subjected to image analysis, which enabled sharpening and visualization of the nanofibers. The first procedure used was the smooth function (first column), thanks to which the amount of noise in the images was significantly reduced. Then, the bandpass filter function (second column) was used, whose task was to filter the image in a banded way. The last function used was the threshold function with the dark back option selected (third column), thanks to which binarization of the images was achieved—only black and white colors appear (the nanofibers are shown in black). The micrographs of the coatings after applying the appropriate procedures are shown in Figure 2.

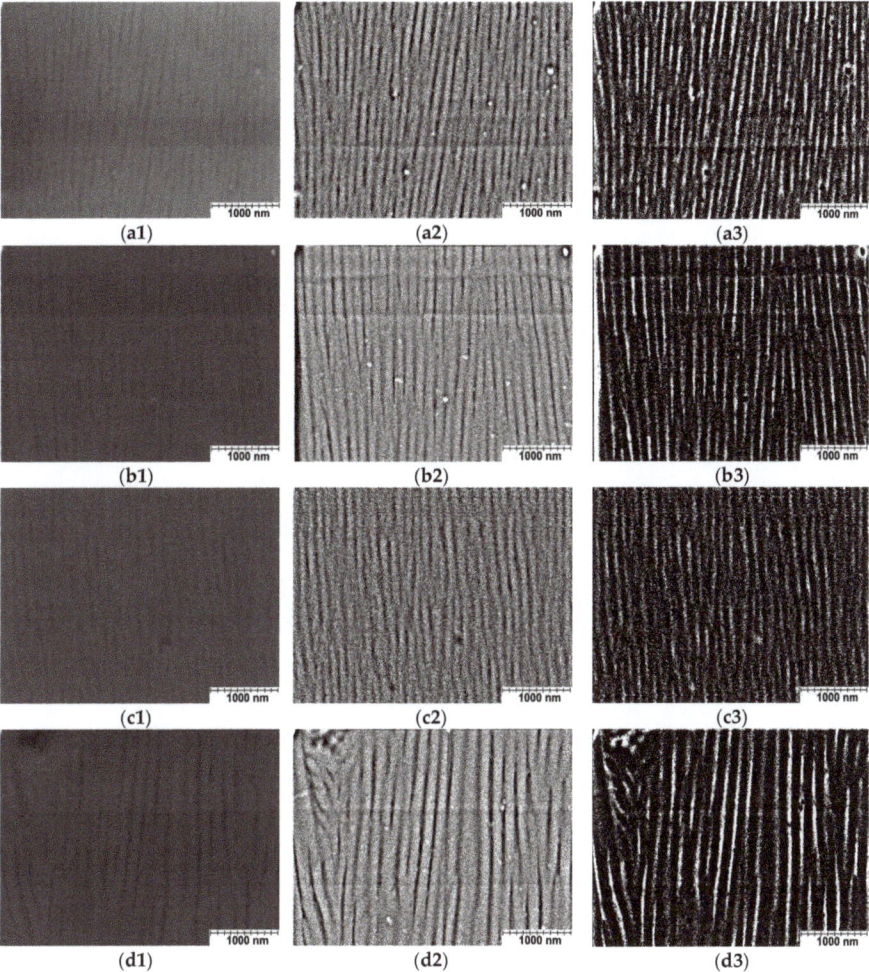

Figure 2. SEM micrograph (cross-section) of nanostructure of the coatings: (**a**) Sample A, (**b**) Sample B, (**c**) Sample C, (**d**) Sample D (sample designations as in Table 2), 1—image with smooth function, 2—image with bandpass filter function, 3—image with threshold function with dark back option selected.

The obtained results of the values of the aluminum oxide fiber diameters and the number of fibers on length 1×10^3 nm are presented in Table 4.

Table 4. Average values of diameters and number of coating fibers.

Sample	Amount of Fibers/nm × 10³	Deviation	Average Fiber Diameter [nm]	Deviation [nm]
A	6.6	0.61	75.99	7.70
B	5.1	0.52	92.52	6.48
C	5.8	0.57	101.87	8.06
D	3.8	0.48	124.59	6.53

The relationship between the coating thickness and the diameter of the aluminum oxide fibers is given by Equation (1).

$$y = 7.83x - 3.8, \tag{1}$$

y—coating thickness,
x—average fibers diameter.

On the basis of the image analysis data obtained from ImageJ 1.50i, histograms of the nanofibers with appropriate diameter groups in individual oxide coatings were made (Figure 3).

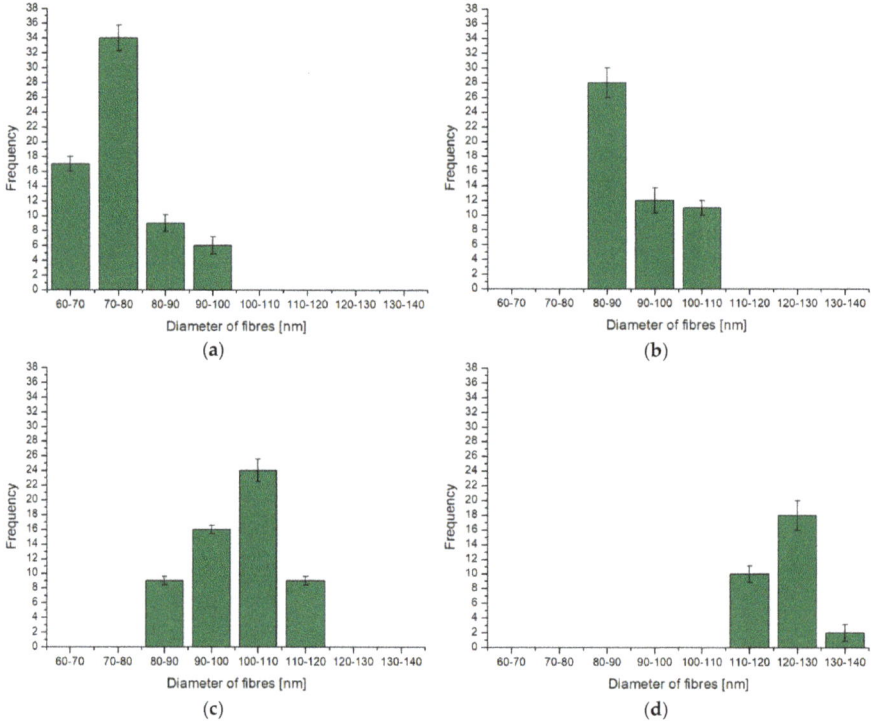

Figure 3. Histograms of nanofibers: (a) Sample A, (b) Sample B, (c) Sample C, (d) Sample D (sample designations in Table 2).

Computer analysis of the coating micrographs showed several significant relationships between the production parameters and the number of nanofibers and their diameters. For a constant current density of 1 A/dm², an increase in electrolyte temperature causes a decrease in the number of nanofibers per 1 × 10³ nm while increasing their diameter (Samples A and B). In turn, at the current density of 4 A/dm² (Samples C and D), the increase in electrolyte temperature causes an increase in the number of nanofibers while reducing the diameter of the fibers. The above relationships most likely result from the increasing secondary solubility of aluminum oxide fibers with increasing temperature at higher current density values. At a constant electrolyte temperature, an increase in current density (Samples A and D) causes a significant reduction in the number of nanofibers while increasing their diameter. Another important relationship is the linear relationship between the average diameter of the Al_2O_3 nanofibers coatings and the thickness of these coatings—an increase in the thickness of the oxide coating increases the diameter of the nanofibers. The coatings produced under extreme conditions (Samples A and D) with the largest and smallest thicknesses are also characterized by the largest and smallest number of nanofibers.

Figure 4 shows the transverse profiles of the surface of the oxide coatings and aluminum alloy made with the profiler after the scratch test. The area above the zero point is marked in green (material upsetting on both sides of the scratch), and in red the area after the scratch was made.

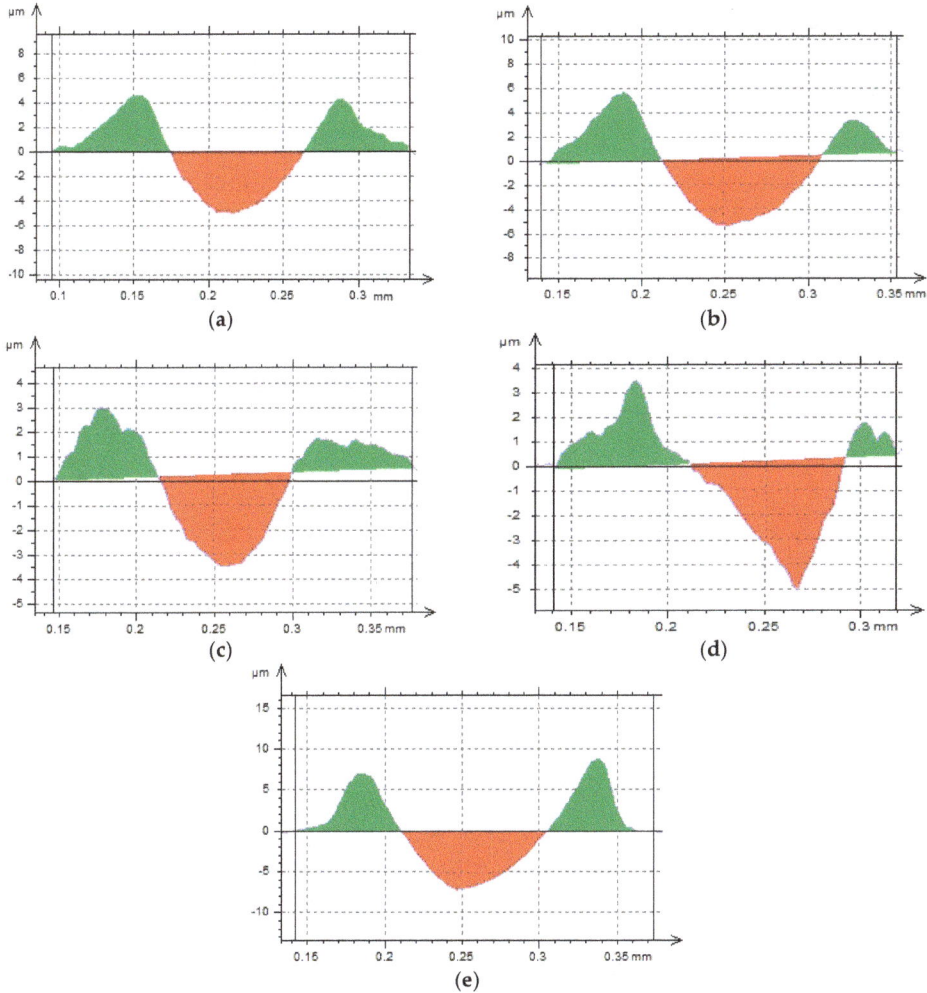

Figure 4. Transverse profiles of oxide coating and aluminum alloy surfaces after scratch test: (**a–d**): Sample designations in Table 2, (**e**): Aluminum alloy EN AW-5251.

The images showing the transverse profiles of the samples show areas both below and above the zero point, whose surface areas change depending on the coating production parameters. The shape of these areas also undergoes change. The aluminum alloy exhibited the most rounded area profiles. The increase in current conditions of the anodizing process leads to "sharpening" of the area profiles, which is caused by the increasing thickness of the oxide coatings and rising hardness. Table 5 shows the values of the ratio of the cross-sectional surface area of material upsetting around Scratch $f1$ and the recess of Scratch $f2$. Based on these values, the processes of anodic alumina coating wear were determined.

Table 5. Values of ration of cross-sectional surface area of material upsetting around Scratch $f1$ and recess of Scratch $f2$.

Sample	$f1/f2$	Deviation $f1/f2$	Wear Process
aluminum alloy	1.085	0.109	grooving
A	1.003	0.083	grooving
B	0.818	0.133	scratching, microcutting
C	0.753	0.230	scratching, microcutting
D	0.481	0.159	scratching, microcutting

Analysis of parameters $f1$ and $f2$ allowed the authors to determine the surface wear process of the tested samples. It was noticed that increasing the thickness of the anodic alumina coating (starting with pure aluminum) reduces the $f1/f2$ dependence. As a result of the analysis, it was found that both in the case of pure aluminum and Sample A (smallest coating thickness) there is only plastic deformation of the surface layer, i.e., grooving. The tested surface is indented by the presence of unevenness or abrasive grain, which causes the material to move outside the surface. Grooving occurs if the value of the ratio $f1/f2 > 1$. In the remaining tested samples, there is scratching and microcutting, which in turn occurs in the case of $0 \leq f1/f2 \leq 1$. Roughness measurements carried out before the scratch test showed slight differences between layer surfaces. The Ra parameter value was 0.35 ± 0.035 μm. Slight differences in roughness do not affect the type of layer wear, however.

4. Conclusions

Based on the conducted research, it was found that the anodizing process parameters of the aluminum alloy EN AW-5251 surface affect the nanostructure of the anodic alumina coatings, whose construction, in turn, affects the type of tribological wear. An increase in current density while maintaining a constant electrolyte temperature results in the creation of an anodic alumina coating with a significantly increased diameter of coating nanofibers, while reducing their quantity. A linear relationship between the increase in the thickness of the oxide coating and the increase in diameter of the nanofibers was also demonstrated. A reduction in the $f1/f2$ ratio value was also observed as the diameter of the nanofibers increased. From among the tested samples, only the surface of the aluminum alloy and anodic alumina coatings of the smallest thickness and the diameter of the nanofibers underwent grooving. The increase in production parameters, and hence the thickness of the coatings causes a change in the wear process of the coating from grooving to scratching and microcutting. Given the structure of the layers (fiber size, coating thickness depending on the density of the electric charge), the above studies are in line with previous scientific studies. Due to the fact that there is no literature research on the impact of nanostructure on the speed and mechanism of wear, we are not able to refer to literature.

Author Contributions: M.B. performed the analysis of the results; M.N. carried out the tests and wrote the manuscript; W.S. contributed to the concept and modified the manuscript. All authors have read and agreed to the published version of the manuscript.

Funding: This research received no external funding.

Conflicts of Interest: The authors declare no conflict of interest.

References

1. Davis, J.R. *Aluminum and Aluminum Alloys*; ASM International: Metals Park, OH, USA, 1993.
2. Davis, J.R. *Corrosion of Aluminum and Aluminum Alloys*; ASM International: Metals Park, OH, USA, 1999.
3. Lee, W.; Park, S.J. Porous anodic aluminum oxide: Anodization and templated synthesis of functional nanostructures. *Chem. Rev.* **2014**, *114*, 7487–7556. [CrossRef] [PubMed]
4. Marcus, P.; Oudar, J. *Corrosion Mechanisms in Theory and Practice*; Marcel Dekker Inc.: New York, NY, USA, 1995.
5. Diggle, J.W.; Downie, T.C.; Goulding, C.W. Anodic oxide films on aluminum. *Chem. Rev.* **1969**, *69*, 365–405. [CrossRef]
6. Runge, J.M. *The Metallurgy of Anodizing Aluminum—Connecting Science to Practice*, 1st ed.; Springer International Publishing: Cham, Switzerland, 2018.
7. Bara, M.; Kmita, T.; Korzekwa, J. Microstructure and properties of composite coatings obtained on aluminium alloys. *Arch. Metall. Mater.* **2016**, *61*, 1107–1112. [CrossRef]
8. Michalska-Domańska, M.; Stępniowski, W.J.; Salerno, M. Effect of inter-electrode separation in the fabrication of nanoporous alumina by anodization. *J. Electroanal. Chem.* **2018**, *823*, 47–53. [CrossRef]
9. Hsing-Hsiang, S.; Shiang-Lin, T. Study of anodic oxidation of aluminum in mixed acid using a pulsed current. *Surf. Coat. Technol.* **2000**, *124*, 278–285.
10. Fratila-Apachitei, L.E.; Duszczyk, J.; Katgerman, L. AlSi(Cu) anodic oxide layers formed in H_2SO_4 at low temperature using different current waveforms. *Surf. Coat. Technol.* **2003**, *165*, 232–240. [CrossRef]
11. Posmyk, A. Co-deposited composite coatings with a ceramic matrix destined for sliding pairs. *Surf. Coat. Technol.* **2012**, *206*, 3342–3349. [CrossRef]
12. Korzekwa, J.; Skoneczny, W.; Dercz, G.; Bara, M. Wear mechanism of Al_2O_3/WS_2 with PEEK/BG plastic. *J. Tribol.* **2014**, *136*, 1–7. [CrossRef]
13. Fratila-Apachitei, L.E.; Tichelaar, F.D.; Thompson, G.E.; Terryn, H.; Skeldon, P.; Duszczyk, J.; Katgerman, L. A transmission electron microscopy study of hard anodic oxide layers on AlSi(Cu) alloys. *Electrochim. Acta* **2004**, *49*, 3169–3177. [CrossRef]
14. Jia, Y.; Zhou, H.; Luo, P.; Luo, S.; Chen, J.; Kuang, Y. Preparation and characteristics of well-aligned macroporous films on aluminum by high voltage anodization in mixed acid. *Surf. Coat. Technol.* **2006**, *201*, 513–518. [CrossRef]
15. Kmita, T.; Bara, M. Surface oxide layers with an increased carbon content for applications in oil-less tribological systems. *Chem. Process Eng.* **2012**, *33*, 479–486. [CrossRef]
16. Kmita, T.; Skoneczny, W. Increase of operational durability of a plastic material-oxide coating couple as a result of the application of a pulsed anodizing process. *Eksploat. I Niezawodn.-Maint. Reliab.* **2010**, *45*, 77–82.
17. Choudhary, R.K.; Mishra, P.; Kaina, V.; Singh, K.; Kumar, S.; Chakravartty, J.K. Scratch behavior of aluminum anodized in oxalic acid: Effect of anodizing potential. *Surf. Coat. Technol.* **2015**, *283*, 135–147. [CrossRef]
18. Xina, Y.; Huoa, K.; Hub, T.; Tanga, G.; Chu, P.K. Mechanical properties of Al_2O_3/Al bi-layer coated AZ91 magnesium alloy. *Thin Solid Films* **2009**, *517*, 5357–5360. [CrossRef]
19. Kmita, T.; Szade, J.; Skoneczny, W. Gradient oxide layers with an increased carbon content on an EN AW-5251 alloy. *Chem. Process Eng.* **2008**, *29*, 375–387.
20. Kubica, M.; Skoneczny, W. The finite element method in tribological studies of polymer materials in tribo-pair with the oxide layer. *Tribol. Lett.* **2013**, *52*, 381–393. [CrossRef]

© 2020 by the authors. Licensee MDPI, Basel, Switzerland. This article is an open access article distributed under the terms and conditions of the Creative Commons Attribution (CC BY) license (http://creativecommons.org/licenses/by/4.0/).

Article

Investigation of the Oxidation Mechanism of Dopamine Functionalization in an AZ31 Magnesium Alloy for Biomedical Applications

Arezoo Ghanbari [1], Fernando Warchomicka [2,*], Christof Sommitsch [2] and Ali Zamanian [1,*]

1. Department of Nano-Technology and Advanced Materials, Institute of Materials and Energy (MERC), 31787-316 Alborz, Iran; ghanbariarezoo91@gmail.com
2. Institute of Materials Science, Joining and Forming, Graz University of Technology, Kopernikusgasse 24/l, A-8010 Graz, Austria; christof.sommitsch@tugraz.at
* Correspondence: fernando.warchomicka@tugraz.at (F.W.); a-zamanian@merc.ac.ir (A.Z.); Tel.: +43-316-873-1654 (F.W.); +98-912-321-1180 (A.Z.)

Received: 9 July 2019; Accepted: 12 September 2019; Published: 16 September 2019

Abstract: Implant design and functionalization are under significant investigation for their ability to enhance bone-implant grafting and, thus, to provide mechanical stability for the device during the healing process. In this area, biomimetic functionalizing polymers like dopamine have been proven to be able to improve the biocompatibility of the material. In this work, the dip coating of dopamine on the surface of the magnesium alloy AZ31 is investigated to determine the effects of oxygen on the functionalization of the material. Two different conditions are applied during the dip coating process: (1) The absence of oxygen in the solution and (2) continuous oxygenation of the solution. Energy dispersive spectroscopy (EDS) and Fourier transform infrared spectroscopy (FTIR) are used to analyze the composition of the formed layers, and the deposition rate on the substrate is determined by molecular dynamic simulation. Electrochemical analysis and cell cultivation are performed to determine the corrosion resistance and cell's behavior, respectively. The high oxygen concentration in the dopamine solution promotes a homogeneous and smooth coating with a drastic increase of the deposition rate. Also, the addition of oxygen into the dip coating process increases the corrosion resistance of the material.

Keywords: biodegradable magnesium; dopamine; Impedance behavior; molecular dynamic simulation

---

1. Introduction

Magnesium implants have intrigued the medical industry as a way to substitute temporary implants with more permanent counterparts. In the application of temporary fasteners, including screws or pins, magnesium implants are superior to permanent titanium or stainless-steel implants, since their use eliminates the need for second surgery to remove the implant after bone healing. Moreover, magnesium alloys provide mechanical properties with close compatibility with bone, thereby decreasing the risk of stress shielding [1,2]. As with other bio metals, magnesium is investigated to promote the cell's response on the surface. The healing rate of bone is critical [3], and, for successful bone regeneration, cell viability is highly significant [4]. Cell adhesion is unlikely to occur directly on the implant. Thus, an appropriate surface treatment is needed [5]. For this treatment, several biomimetic coatings are investigated: Aminopropyltriehtoxysilane, vitamin C (AV), carbonyldiimidazole (CDI), stearic acid (SA), aminated hydroxyethyl cellulose, gelatin, and peptides [5–9]. Depending on the composition, biomimetic coatings can have the capability to retard the corrosion affinity of magnesium to body fluids [5,8,9]. Therefore, a biomimetic coating should be adhesive and compact enough to protect the substrate against corrosive body fluid. In this area of study, polydopamine is introduced as a biomimetic coating to enhance the adhesion of other biomimetic coatings [10]. For instance, the use of a coating made by gelatin (a mixture of peptides and proteins) and modified with polydopamine can improve the binding affinity between the coating and the metal (e.g., on a titanium surface [11]) or promote corrosion resistance [12]. In other cases, the use of dopamine helps to immobilize functional molecules as a robust anchor in biofunctional surfaces (e.g., nano magnetic particles [13]) or increase the cytocompatibility and osteogenic potential of the scaffold [14], among other examples.

Recent works on magnesium have shown the advantage of using a functionalized surface with dopamine to enhance corrosion resistance [15–17], as well as the cell proliferation on the surface of the implant [17,18]. Inspired by the composition of adhesive proteins in mussels, dopamine can be self-polymerized to form a surface-adherent, thin, and robust film on almost all kinds of materials, including organic and inorganic ones [19]. In this area, the dip coating method [15–17,20] has been used as a convenient way to anchor the dopamine and create a film on magnesium alloys. In situ spontaneous deposition takes place through an aerobic auto-oxidation mechanism in a mildly basic aqueous buffered media and, consequently, a bio-inspired film is generated on the substrate. Although the film is known as polydopamine, there is controversy over whether the film consists of supramolecular of monomer aggregates or has a polymeric structure [21]. The oxidation mechanism during the dopamine coating and the evolution of the molecules have been deeply investigated [18,20,22–25]. Many oxidants were evaluated as a buffer for the formation of the coating, including compounds like copper sulfate, sodium periodate, and ammonium peroxodisulfate [25,26], and oxygen as a biocompatible oxidant [20,23,25,26]. It is reported that polydopamine or dopamine- melanin film deposition seems to depend not only on the monomer's availability but also on the availability of O_2 [16,25].

Whether polydopamine is incorporated in a magnesium implant as an adhesive intermediate biomimetic layer or an individual biomimetic layer, the corrosion rate of this layer is fundamental. The present work focuses the investigation on variation of the corrosion resistance of the AZ31 alloy through functionalization of the surface via dopamine, with the presence and absence of oxygen. Oxygen is selected as a condition variable to investigate the effects of an oxidant on the coating's quality.

2. Materials and Methods

2.1. Substrate

The magnesium alloy AZ31 was used in this work. The chemical composition of AZ31, in wt.%, is as follows: 3 Al, 0.4 Mn, 0.8 Zn, max. 0.005 Fe, 0.004 Cu, 0.003 Ni, 0.07 Si, max. 0.01 Cr, and balance Mg. The as-received AZ31 was in a cold rolled and partially annealed condition.

2.2. Coating Process

Sheet samples of 1×1 cm^2 were polished on both sides with SiC abrasive papers (up to 4000 grit) to obtain a uniform roughness. Samples were washed in acetone for 10 min in an ultrasonic bath to remove any surface contaminations. At the end of the surface preparation, samples were washed in deionized water and immediately dried in the air.

For the dip coating, dopamine hydrochloride (($HO)_2C_6H_3CH_2CH_2NH_2 \cdot HCL$), hydrochloric acid (HCl), and tris(hydroxymethyl)aminomethane, also known as Tris, (($HOCH_2)_3CNH_2$) were purchased from Merck (Darmstadt, Germany) as raw materials. The coating solution was prepared using 2 mg/mL dopamine hydrochloride, 100 mL deionized water, and 48 mM Tris as a buffer. In order to understand the oxidant effect on the coating, two different conditions were applied during the dip coating process: (1) the absence of oxygen in the solution and (2) the solution continuously being oxygenized with a flow rate of 0.4 L/min. In both conditions, the pH of the solution was adjusted to 8.5 and was kept constant by adding adequate amounts of hydrochloric acid with a concentration of 1 M. The pH value was monitored with a digital pH meter (Metrohm, Herisau, Switzerland).

The substrate was immersed in the solution for two hours in darkness [17,27]. After immersion, the samples were removed quickly to inhibit further oxidation and cleaned in deionized water to remove any unattached dopamine on the surface. Coated samples were finally dried in a nitrogen flow to avoid any undesirable surface oxidation. In this work, samples functionalized via dopamine with the absence of oxygen are designated as "dopa", while samples immersed in a solution rich in oxygen are named "dopa-O", and the pristine substrate is "AZ31".

2.3. Characterization

The morphological feature observations, accompanied with a chemical analysis of the pristine and functionalized AZ31 alloys, were carried out using a field emission scanning electron microscope (FESEM, Mira3 Tescan, Brno, Czech Republic) equipped with an energy dispersive spectroscope detector (EDS: EDAX- AMETEK, Tilburg, The Netherlands). FESEM images were taken in secondary electron mode to monitor the topography of the layer. Chemical analysis was carried at the surface and at the cross section to estimate the homogeneity of the deposited polydopamine. Moreover, Fourier transform infrared spectroscopy (FTIR, Thermo; Waltham, MA, USA) in transmittance mode was used to verify the presence of polymerized dopamine on the surface of the "dopa" and "dopa-O" samples.

2.4. Electrochemical Analysis

The degradation behavior of the samples was investigated via electrochemical impedance spectroscopy (EIS; AutoLab Potentiostat, Metrohm, Herisau, Switzerland). Samples were stabilized in phosphate-buffered saline (PBS) solution prior to the tests, and all impedance measurements were performed in a PBS solution at a temperature of 37 °C. A three-electrode electrochemical cell setup, with an Ag/AgCl electrode as a reference electrode and a platinum sheet as a counter electrode, was utilized in this work. Moreover, samples were used as working electrodes. EIS scans were acquired from 100 kHz to 0.01 Hz, and the impedance data were analyzed with the NOVA software (version 1.11).

2.5. Cell Cultivation

A human osteosarcoma cell line (G292), taken from National Cell Bank of Iran, was selected to evaluate the cellular behavior of samples. Osteoblast-like cells were cultured in a humidified atmosphere of 5% CO_2 at a temperature of 37 °C using Dulbecco's modified Eagle's medium (DMEM, BIO-IDEA, Iran) supplemented with 10% fetal bovine serum (FBS) (Gibco, Thermo, Waltham, MA, USA) and 1% antimicrobial penicillin/streptomycin solution 100X (Biosera, Kansas City, MO, USA). The cells were subcultured in a cell culture medium, which was refreshed every 3 days before utilization. Two series of samples were placed in 6-well plates and sterilized by ultraviolet light in 70% (V/V) ethanol–water solution and kept in sterilized media before analysis. Afterwards, samples were seeded with the same number of G292 cells (10^5 cells) in a culture medium, and the cells were cultivated for 72 h in an incubator. Then, the samples were gently rinsed with phosphate-buffered saline solution (PBS) to remove the unattached cells.

Microscopic observations were carried out via both FESEM and fluorescent microscope to observe the cell morphology and adhesion to the samples. For the FESEM observation, the cells were fixed with a 2.5 v % glutaraldehyde solution for 2 h in a dark environment. Subsequently, the samples were gently washed again with PBS and then immersed in a bath of 50%, 60%, 70%, 80%, 90%, and 100% (V/V) ethanol / water solutions for 30 min each time to dehydrate the cells attached to their surfaces. The dehydration conducted in absolute ethanol was performed twice, and each time the samples were immersed in 100% ethanol for 1 h. The morphology of the G292 cells attached to the samples was observed after drying and sputtering with gold.

The cell behavior was also examined by a fluorescence microscope (Olympus, Japan) using a cell staining method. A total of 0.1% acridine orange (AO, sigma, Saint Louis, MO, USA) was utilized to visualize the remaining cells on the substrates [28], staining the cells for 1–2 min.

2.6. Molecular Dynamics Simulation

The effect of the oxidant on polydopamine coating was studied via molecular dynamics simulations. Two different conditions were designed, simulating the "dopa" (condition one) and "dopa-O" (condition 2) deposition conditions. In both conditions, the magnesium matrix was simulated via EAM (embedded atom model) potential. Moreover, each condition consisted of a mixture of dopamine monomers, indolequinone monomers, and dimmers. Concerning the oxidation mechanism of dopamine, the different ratios of dopamine to indolequinone were considered for each deposition condition: 90/10 for "dopa" and 10/90 for "dopa-O". Molecular dynamics simulations were performed via a large-scale atomic/molecular massively parallel simulator (LAMMPS). The atomic computations were carried out in periodic boundaries and in three dimensions. The generic force field, Dreiding, was employed to describe the inter-molecular interactions of the polymer. In this case, the valence interactions were considered to be bond stretch, bond-angle bend, and dihedral angle torsion terms [29]. The non-bonded or Van der Waals interactions of polymer/polymer and polymer/Mg are given by the Lennard–Jones [30] potential. A total of 10,000,000 time-steps with an increment of 0.01 fs were simulated for each condition. The temperatures of the conditions were controlled using a Langevin thermostat accompanied by a micro-canonical ensemble NVE. Visualizations were created using the Ovito software.

3. Results

3.1. Characterization of the Coating

The surface observation and the elemental map analysis of "AZ31", "dopa", and "dopa-O" conditions are displayed in Figure 1. In the as-received "AZ31" condition, intermetallic phases rich in aluminum are observed on the polished surface, which is reported as a typical secondary phase $Mg_{17}Al_{12}$ in AZ31 alloy [31,32]. Figure 1b, c shows the functionalized surface of the "dopa" and "dopa-O". Both samples are homogeneously covered with a coating, which exhibits the presence of nitrogen and oxygen on the Mg matrix, in concordance with observations in the literature, together with carbon and hydrogen elements in the polydopamine layer [24]. At the surface of the coating, cracks are formed during the drying stage [33].

A cross section of the coated samples is analyzed in Figure 2 to determine the thickness of the layer, and the variation of the elements depends on the dip condition. The thickness of the coatings was 42 and 85 μm for "dopa" and "dopa-O", respectively, in agreement with both SEM observations and element profile scans. The element scan line shows a strong influence of the type of coating on the amount of nitrogen and oxygen elements in the layers. The "dopa-O" has higher intensity than the "dopa" condition. Furthermore, the presence of Mg in the layer of "dopa-O" may be related to the oxidation of the substrate due to the flow of oxygen during the coating process. The element profile in Figure 2c shows that although the amount of element is lower in "dopa", it is more homogenous in composition than that the "dopa-O". The first 20 μm (Region I) keeps a constant composition for both conditions, while the reduction of N through the coating in "dopa-O" (region II) indicates a lack of polydopamine in this area, which has been replaced by magnesium oxide.

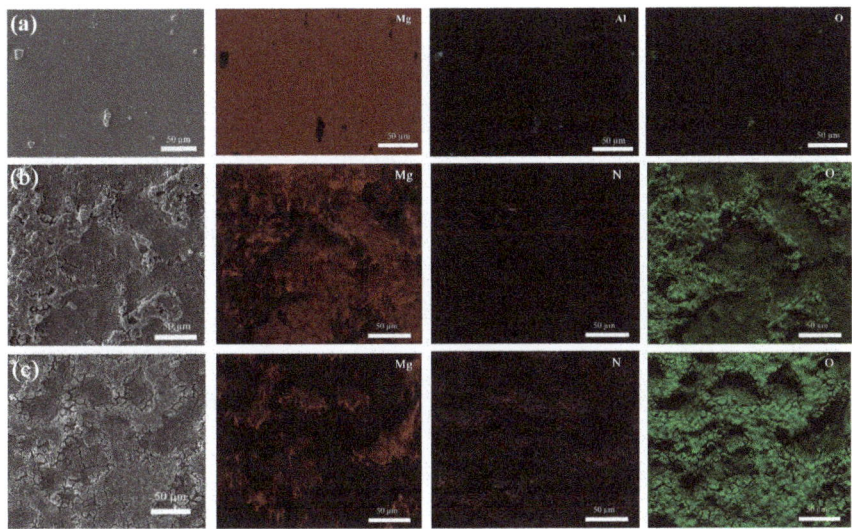

Figure 1. The SEM micrographs and relative elemental mapping of (**a**) the as-received AZ31 alloy; (**b**) "dopa" and (**c**) "dopa-O". Samples functionalized via dopamine with the absence of oxygen are designated as "dopa", while samples immersed in a solution rich in oxygen are named "dopa-O".

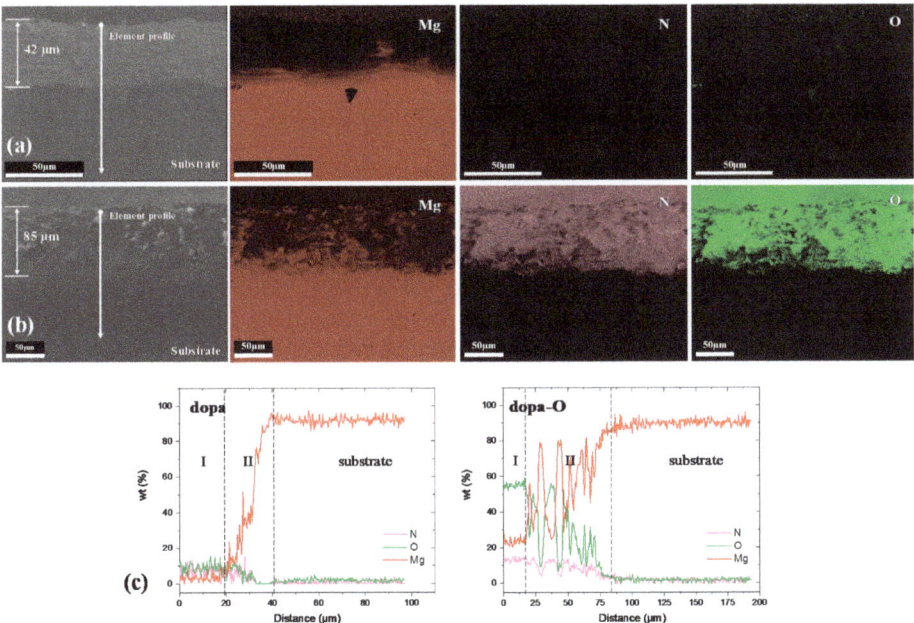

Figure 2. SEM micrographs and relative elemental mapping of the cross section (**a**) "dopa" and (**b**) "dopa-O"; (**c**) elemental line scan of "dopa" and "dopa-O" coatings, marked in the micrographs in (**a**) and (**b**), respectively.

FTIR characterizations detect the functional groups in the polydopamine layer. The functional groups of the formed polymeric layer on the surface of "dopa" and "dopa-O" are depicted in Figure 3. A peak at 702 cm^{-1} can be observed in the "dopa" curve, which implies the ring deformation vibration of 1,2,4-trisubstituted benzene. In both conditions, the peaks at 800 cm^{-1} (and also the band broadening between 800 and 860 cm^{-1}) can be attributed to the C–H wagging vibration of the ring hydrogens in the 1,2,4-trisubstituted and 1,2,4,5-tetrasubstituted benzenes. The band accompanied with peaks in the range of 950–1420 cm^{-1} is ascribed to ring deformation vibrations and C–N stretching vibrations [34]. The adsorption bands located in the range of 1450–1625 cm^{-1} originate from the C=C stretching vibration of benzene rings and the N–H bending [27]. In addition, a peak at 1637 for both conditions is attributed to the C=C stretching vibrations, while the "dopa-O" condition shows a peak at 1729 cm^{-1} related to the C=O stretching vibrations [34]. A broad absorbance in the spectrum between 3100 and 3600 cm^{-1} is ascribed to N–H/O–H stretching vibrations [33], and in this region, NH and NH$_2$ stretching vibrations happen in the range of 3100–3360 cm^{-1} [34].

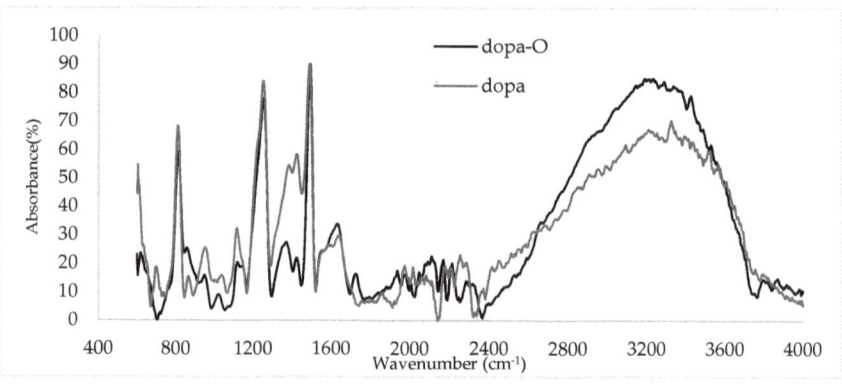

Figure 3. The Fourier transform infrared spectroscopy (FTIR) spectrum of "dopa" and "dopa-O".

3.2. Electrochemical Properties

The effect of oxygen on the corrosion resistance of the coated samples was analyzed by impedance tests in a PBS solution at 37 °C. The variation of the real impedance component (Z') versus the imaginary impedance component (Z''), the Nyquist plot, is shown in Figure 4a. The Bode plots (given in Figure 4c,d) improve the accuracy of the degradation behavior. Herein, the consistency of the obtained EIS data are evaluated with the Kramers–Kronig (K–K) transformation. The impedance spectra of all samples are characterized in high and medium frequency regions, with one semicircle loop in the Nyquist plot. Moreover, the one phase extremum in the Bode plot (Figure 4c) is an indicator of one semicircle in the Nyquist plot (Figure 4a). Figure 4d illustrates the decrease in impedance when the frequency decreases. This process is influenced by the adsorbed intermediate species, which form layers of corrosion products [35].

To compare the impedance data of the samples in quantity, the impedance plots are fitted with an equivalent electrical circuit, which is depicted in Figure 4b. The semicircle in the high and medium frequency region is ascribed as the constant phase element (*CPE*), simulating the charge transfer resistance (R_c) in parallel with the electric double layer. Moreover, the inductive loop is simulated by the inductance (*L*) and the relative inductive resistance (R_l). The total circuit is in series with the solution resistance (R_s).

The impedance data extracted from the equivalent electrical circuit are given in Table 1. From the data, the quantity of the circuit elements is increased in the following order: "dopa-O" > AZ31 > "dopa". Due to the insufficient quality of the polymer coating, the impedance is decreased in the "dopa". However, the "dopa-O" sample shows the highest constant phase element, inductance, and real resistances. Apparently, the presence of oxygen gas flowing during the dopamine dip coating process improves the corrosion resistance of the functionalized AZ31 alloy.

Table 1. The extracted impedance data from equivalent electrical circuit. CPE, constant phase element.

Sample	R_s (Ω cm^2)	R (Ω cm^2)	CPE (μ$_{Mho}$/cm^2)	N	R_l (Ω cm^2)	L (H cm^2)	X^2 (<0.5)
AZ31	22.1	68.7	39.6	0.841	38.3	13.9	0.385
dopa	13.152	31.44	50.83	0.652	15.408	71.52	0.427
dopa-O	15.38	142.32	105.83	0.718	185.52	264	0.206

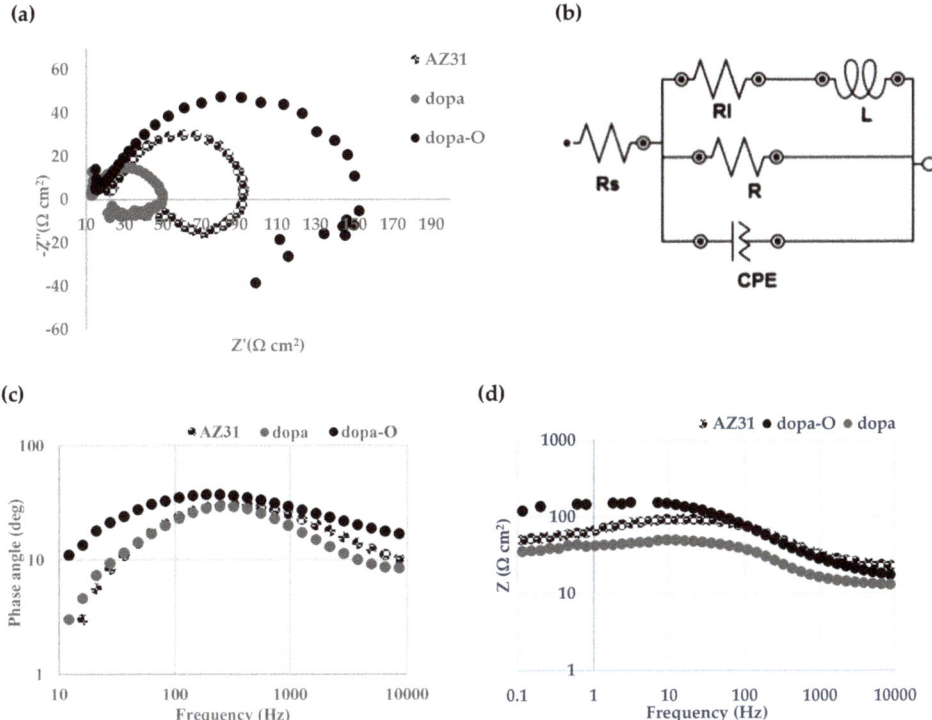

Figure 4. Electrochemical data for samples; "AZ31", "dopa", and "dopa-O" (**a**) Nyquist plot; (**b**) Equivalent electrical circuit; (**c**) Bode-phase angle plot, and (**d**) Bode-impedance plot.

3.3. Cell Cultivation

The SEM micrographs of the G292 cells attached to the samples, after 72 h cultivation, are illustrated in Figure 5. Artifacts observed on the surface might be related to the sputtering with gold. The cells keep a rounded and spherical shape and show a slightly tendency to be spread out on the surface of the "AZ31" alloy and "dopa" (Figure 5a,b). However, the cells also show a tendency to spread out on the surface of the "dopa-O" condition, as illustrated in Figure 5c.

Figure 5. SEM micrographs of the adherent G292 cells, (**a**) "AZ31"; (**b**) "dopa"; and (**c**) "dopa-O" after 72 h of cultivation time.

Figure 6 shows the G292 cells attached on the surface of samples by fluorescent microscopy. The adhesion of the cells can be evidenced in all the cases and a change of the cell's behavior in the "dopa-O" condition is perceptible. However, this information is not enough to draw the effect of the coating on the cell's morphology due to the cytoskeleton/membrane cannot be observed properly. Further investigations focused on cell viability and proliferation are needed to better quantify the effect of the oxygen on the cells' behavior.

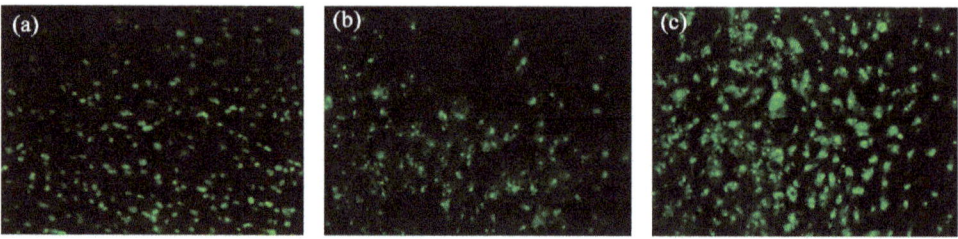

Figure 6. Fluorescent microscopy of adherent cells stained with acridine orange: (**a**) "AZ31"; (**b**) "dopa"; and (**c**) "dopa-O" after 24 h of cultivation time; (Magnification 100×).

3.4. Molecular Dynamics Simulation

Molecular dynamics simulations help to determine the effect of oxidants on the polydopamine of the AZ31 alloy substrate. The displacement of the polymer mass towards the substrate surface was simulated. Thus, the center-of-mass (com) command was computed in LAMMPS for Z direction. Additionally, two different conditions were implemented by LAMMPS, considering that the dopamine molecule transforms into indolequinone in the presence of oxygen [25], as described in Section 2.6. The dimmers were designed for both conditions due to the further interactions between the monomer molecules [36]. In fact, polydopamine is not a covalent bond polymer but instead an aggregate of monomers held together by noncovalent forces [24]. The designed molecular structures of dopamine and indolequinone monomers [25], along with the dimmers [24], are illustrated in Figure 7a–c. In addition, the visualized conditions (one and two) are illustrated in Figure 7d,e. The displacements of the center-of-mass (com) in the polymer were calculated for each time-step in the Z direction, shown in Figure 7f. The slope of the graphs represents the difference of the displacement per difference of time. Hence, the slope indicates the deposition rate (angstrom/femtosecond). In this work, the slope of the "dopa-O" condition is higher, showing that the presence of oxygen accelerates the deposition of the polymer on the substrate. It seems that the tendency of indolequinone to bond to the substrate is higher than that of dopamine during the coating process.

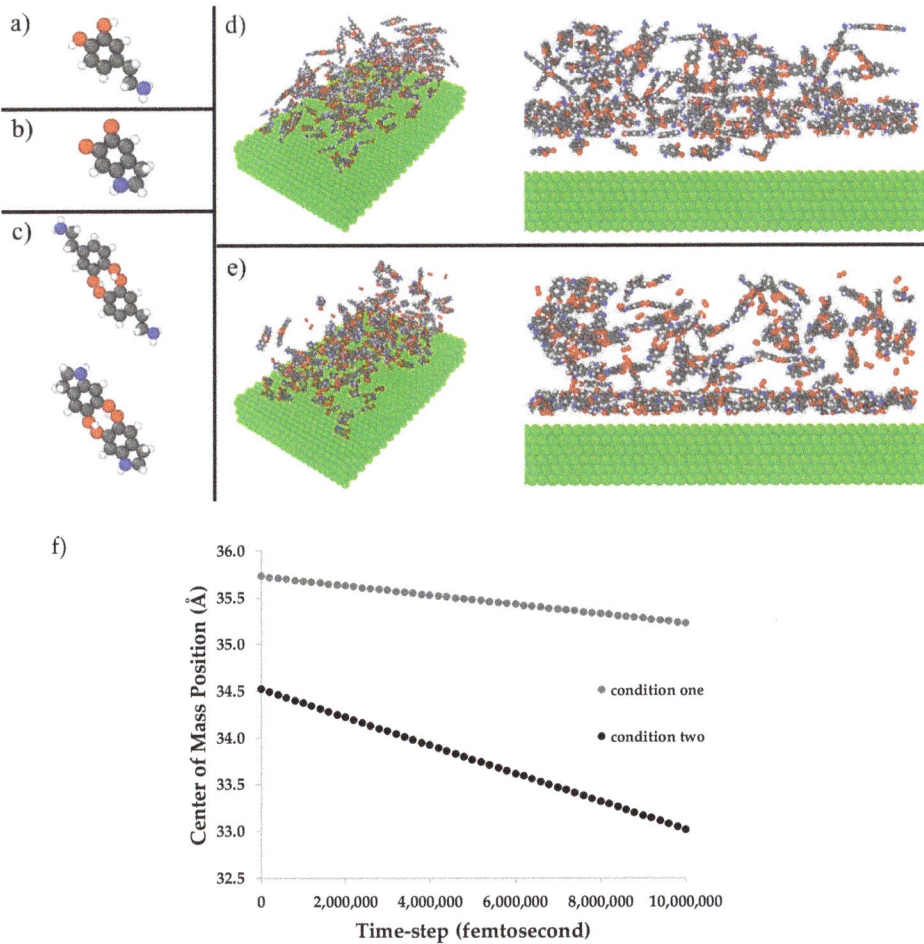

Figure 7. Molecular dynamic simulation: (**a**) Dopamine monomer and (**b**) indolequinone monomer; (**c**) dimers; (**d**) condition one; (**e**) condition two; (colors definition: green: Mg; gray: C; white: H; red: O and blue: N); (**f**) center of mass versus time-step.

4. Discussion

4.1. Characterization of the Coating

The oxidant can intensify polydopamine formation [18,25]. According to the EDS maps given in Figure 1, the surface of the "dopa" and "dopa-O" is covered homogeneously with oxygen and nitrogen. The thickness measurements given in Figure 2 indicate that the thickness of the coating layer in "dopa-O" is, relatively, twice that of "dopa", in agreement with the deposition rate estimated by molecular dynamic simulations (Figure 7f). From respective cross section EDS mapping, the difference in nitrogen and oxygen content is obvious. This difference can be further perceived from the element profile scans (Figure 2c). The coatings (in depth) show two main regions, where the depth of region I is relatively the same for both conditions (~20 μm). Region II shows more depth for the "dopa-O", and the presence of Mg in both layers might be related to the magnesium oxide formed during the immersion in the dip process. This effect is more noticeable in the "dopa-O" condition, where the continuous flowing of gas can promote the formation and growth of the oxide at the surface. On the other hand, a difference between the two conditions with respect to the ratio of oxygen to nitrogen in weight percent (O/N) is pronounced. Region I indicates a ratio of around 3 for "dopa-O" and 1 for "dopa". The presence of polydopamine should show a ratio of O/N in dopamine and an indolequinone ratio of 2. For this reason, the presence of these molecules on the surface of "dopa" is more reliable in this work. Analysis of the polydopamine layer via FTIR spectrum (Figure 3) indicated that the NH and NH_2 stretching vibrations take place in the wavenumber range, 3100–3360 cm^{-1} [34], and a higher absorbance of "dopa-O" is associated with the amount of nitrogen (Figure 2c). Through the oxidation mechanism of polydopamine, the dopamine molecules transform to indolequinone molecules, whose structure is benzene-like with different positions for the substitutes. Here, the three-substituted structure of dopamine changes to the four-substituted structure of indolequinone [25]. The peak at 702 cm^{-1} in "dopa" is due to the ring deformation vibration of 1,2,4-trisubstituted benzene, which is similar to the dopamine structure [34]. Thus, a high amount of dopamine that is not fully oxidized can still be expected in the coating of the "dopa". If oxidation and cyclization do not take place completely, dopamine molecules co-exist in the solution [37].

On the other hand, the peak at 800 cm^{-1} and the band broadening between 800 and 860 cm^{-1} are attributed to the C–H wagging vibration of ring hydrogens in 1,2,4-trisubstituted and 1,2,4,5-tetrasubstituted benzenes [34]. This peak and band are slightly higher in the "dopa-O" condition, which has a similar structure to 1,2,4,5-tettrasubstituted benzenes. This means that the presence of oxygen activates the oxidation mechanism, and more three-substituted benzenes (dopamine structures) transform into four-substituted benzenes (indolequinone structures). Moreover, the appearance of a peak at 1729 cm^{-1} for "dopa-O" is due to the carbonyl (C=O) stretching vibrations of the indolequinone [24,25,34] structure, showing that adding oxygen increases the ratio of the dopamine that transforms to indolequinone. This effect promotes the presence of polydopamine instead of dopamine molecules in the "dopa-O" condition.

4.2. Electrochemical Properties

A comparison of the impedance responses to the corrosive PBS solution (Figure 4a) shows that the largest semicircle is in the "dopa-O" condition, thereby validating the observations of Kim et al. in various commercial microporous membranes [38]. The amount of the constant phase element, representing the resistance to the adsorption/diffusion processes at the electrode surface [39], increased from 50.83 μ_{Mho}/cm^2 in "dopa" to 105.83 μ_{Mho}/cm^2 in "dopa-O". Thus, the impedance of the capacity behavior (CPE) is relatively doubled (see Table 1). This trend is also observed for real resistors. The charge transfer resistance increased from 31.44 Ω cm^2 in "dopa" to 142.32 Ω cm^2 in "dopa-O". Thus, the transferring of charges responsible for anodic and cathodic reactions is hindered in "dopa-O", so the formed polymeric layer on "dopa-O" acts as a stronger inhibitor of charge transfer and charge storage (CPE) than "dopa". Moreover, there is a considerable increase in inductance element values after surface functionalization in the presence of oxygen, representing greater prevention against specious movement, including Mg^{2+} and $Mg(OH)_2$. On the other hand, the equivalent circuit component values of "dopa" are lower than those of "AZ31". It seems that the polymeric layer formed on the surface of "dopa" cannot act as an effective barrier to magnesium dissolution in the corrosive PBS solution.

The polymeric layer formed in the "dopa-O" deposition condition acts as a superior barrier layer to PBS corrosive media. As seen in the EDS line (Figure 2c), region I in "dopa-O" contains a higher O/N ratio. Further, according to the FTIR view, the functional layer groups indicate a greater oxidation state. Since oxidation is the polymerization mechanism, it can be deduced that the polymerization of dopamine takes place more frequently on "dopa-O". Through dopamine polymerization or polydopamine formation, aggregates are formed on the surface by linking the particles together [33]. Therefore, this layer can act as a better barrier than "dopa" against corrosive liquid penetration, with less oxidized dopamine in the layer. On the other hand, the presence of MgO in "dopa-O" might influence the corrosion resistance of the material.

5. Conclusions

The use of oxygen as an oxidant agent for the dopamine functionalizing process was considered in this work, using the magnesium alloy AZ31. The introduction of oxygen into the dip coating draws the following conclusions:

- The molecular dynamic simulation and FTIR analysis show an improvement in the deposition rate and the presence of polydopamine on the substrate.
- The addition of oxygen increases the impedance response of the AZ31 alloy.
- The formation of polydopamine with the presence of amino and hydroxyl functional groups on the surface of the material might be promising for the cell's response. Further investigations focused on the cell viability and proliferation are necessary to confirm the biocompatibility of the coating.

Author Contributions: Conceptualization, F.W., A.G. and A.Z.; methodology, A.Z.; software, A.G.; formal analysis, A.G. and F.W.; investigation, F.W. and A.G.; writing—original draft preparation, A.G., F.W. and A.Z.; writing—review and editing, A.G., F.W., C.S. and A.Z.; supervision, A.Z. and C.S.

Funding: This research received no external funding.

Acknowledgments: Open Access Funding by the Graz University of Technology. Authors would like to thank R. Buzolin for FESEM observations, A. Bordbar-Khiabani for samples preparation, T. Ramezani for helping with cell cultivation and R. Sabetvand for supporting the molecular dynamics simulations.

Conflicts of Interest: The authors declare no conflict of interest.

References

1. Witte, F.; Hort, N.; Vogt, C.; Cohen, S.; Kainer, K.U.; Willumeit, R.; Feyerabend, F. Degradable biomaterials based on magnesium corrosion. *Curr. Opin. Solid State Mater. Sci.* **2008**, *12*, 63–72. [CrossRef]
2. Gu, X.N.; Zheng, Y.F. A review on magnesium alloys as biodegradable materials. *Front. Mater. Sci. Chin.* **2010**, *4*, 111–115. [CrossRef]
3. Ghaffarpasand, F.; Shahrezaei, M.; Dehghankhalili, M. Effects of platelet rich plasma on healing rate of long bone non-union fractures: A randomized double-blind placebo controlled clinical trial. *Bull. Emerg. Trauma* **2016**, *4*, 134–140. [PubMed]
4. Shahrezaee, M.; Salehi, M.; Keshtkari, S.; Oryan, A.; Kamali, A.; Shekarchi, B. In vitro and in vivo investigation of pla/pcl scaffold coated with metformin-loaded gelatin nanocarriers in regeneration of critical-sized bone defects. *Nanomedicine* **2018**, *14*, 2061–2073. [CrossRef] [PubMed]
5. Wagener, V.; Schilling, A.; Mainka, A.; Hennig, D.; Gerum, R.; Kelch, M.-L.; Keim, S.; Fabry, B.; Virtanen, S. Cell adhesion on surface-functionalized magnesium. *ACS Appl. Mater. Interfaces* **2016**, *8*, 11998–12006. [CrossRef] [PubMed]
6. Wagener, V.; Killian, M.S.; Turhan, C.M.; Virtanen, S. Albumin coating on magnesium via linker molecules—comparing different coating mechanisms. *Colloids Surf. B Biointerfaces* **2013**, *103*, 586–594. [CrossRef] [PubMed]
7. Cui, W.; Beniash, E.; Gawalt, E.; Xu, Z.; Sfeir, C. Biomimetic coating of magnesium alloy for enhanced corrosion resistance and calcium phosphate deposition. *Acta Biomater.* **2013**, *9*, 8650–8659. [CrossRef] [PubMed]
8. Zhu, B.; Xu, Y.; Sun, J.; Yang, L.; Guo, C.; Liang, J.; Cao, B. Preparation and characterization of aminated hydroxyethyl cellulose-induced biomimetic hydroxyapatite coatings on the AZ31 magnesium alloy. *Metals* **2017**, *7*, 214. [CrossRef]
9. Tiyyagura, H.R.; Fuchs-Godec, R.; Gorgieva, S.; Arthanari, S.; Mohan, M.K.; Kokol, V. Biomimetic gelatine coating for less-corrosive and surface bioactive Mg–9Al–1Zn alloys. *J. Mater. Res.* **2018**, *33*, 1449–1462. [CrossRef]
10. Lee, H.; Rho, J.; Messersmith, P.B. Facile conjugation of biomolecules onto surfaces via mussel adhesive protein inspired coatings. *Adv. Mater.* **2009**, *21*, 431–434. [CrossRef] [PubMed]

11. Yang, X.; Zhu, L.; Tada, S.; Zhou, D.; Kitajima, T.; Isoshima, T.; Yoshida, Y.; Nakamura, M.; Yan, W.; Ito, Y. Mussel-inspired human gelatin nanocoating for creating biologically adhesive surfaces. *Int. J. Nanomed.* **2014**, *9*, 2753–2765.
12. Zhou, X.; Ouyang, J.; Li, L.; Liu, Q.; Liu, C.; Tang, M.; Deng, Y.; Lei, T. In vitro and in vivo anti-corrosion properties and biocompatibility of 5β-TCP/Mg-3Zn scaffold coated with dopamine-gelatin composite. *Surf. Coat. Technol.* **2019**, *374*, 152–163. [CrossRef]
13. Xu, C.; Xu, K.; Gu, H.; Zheng, R.; Liu, H.; Zhang, X.; Guo, Z.; Xu, B. Dopamine as a robust anchor to immobilize functional molecules on the iron oxide shell of magnetic nanoparticles. *J. Am. Chem. Soc.* **2004**, *126*, 9938–9939. [CrossRef] [PubMed]
14. Liu, Y.; Xu, C.; Gu, Y.; Shen, X.; Zhang, Y.; Lie, B.; Chen, L. Polydopamine-modified poly(l-lactic acid) nanofiber scaffolds immobilized with an osteogenic growth peptide for bone tissue regeneration. *R. Soc. Chem. Rsc Adv.* **2019**, *9*, 11722–11736. [CrossRef]
15. Huang, L.; Yi, J.; Gao, Q.; Wang, X.; Chen, Y.; Liu, P. Carboxymethyl chitosan functionalization of cped-treated magnesium alloy via polydopamine as intermediate layer. *Surf. Coat. Technol.* **2014**, *258*, 664–671. [CrossRef]
16. Singer, F.; Schlesak, M.; Mebert, C.; Höhn, S.; Virtanen, S. Corrosion properties of polydopamine coatings formed in one-step immersion process on magnesium. *Appl. Mater. Interfaces* **2015**, *7*, 26758–26766. [CrossRef]
17. Lin, B.; Zhong, M.; Zheng, C.; Cao, L.; Wang, D.; Wang, L.; Liang, J.; Cao, B. Preparation and characterization of dopamine-induced biomimetic hydroxyapatite coatings on the AZ31 magnesium alloy. *Surf. Coat. Technol.* **2015**, *281*, 82–88. [CrossRef]
18. Lynge, M.E.; Westen, R.V.D.; Postmab, A.; Stadler, B. Polydopamine—A nature-inspired polymer coating for biomedical science. *Nanoscale* **2011**, *3*, 4916–4928. [CrossRef]
19. Lee, H.; Dellatore, S.M.; Miller, W.M.; Messersmith, P.B. Mussel-inspired surface chemistry for multifunctional coatings. *Science* **2007**, *318*, 426–430. [CrossRef]
20. Yang, H.-C.; Wu, Q.-Y.; Wan, L.-S.; Xu, Z.-K. Polydopamine gradients by oxygen diffusion controlled autoxidation. *Chem. Commun.* **2013**, *49*, 10522–10524. [CrossRef]
21. Sedó, J.; Saiz-Poseu, J.; Busqué, F.; Ruiz-Molina, D. Catechol-based biomimetic functional materials. *Adv. Mater.* **2013**, *25*, 653–701. [CrossRef] [PubMed]
22. Hu, H.; Dyke, J.C.; Bowman, B.A.; Ko, C.-C.; You, W. Investigation of dopamine analogues: Synthesis, mechanistic understanding and structure-property relationship. *Langmuir* **2016**, *32*, 9873–9882. [CrossRef] [PubMed]
23. Herlinger, E.; Jameson, R.F.; Linert, W. Spontaneous autoxidation of dopamine. *J. Chem. Soc. Perkin Trans.* **1995**, *2*, 259–263. [CrossRef]
24. Dreyer, D.R.; Miller, D.J.; Freeman, B.D.; Paul, D.R.; Bielawski, C.W. Elucidating the structure of poly(dopamine). *Langmuir* **2012**, *28*, 6428–6435. [CrossRef] [PubMed]
25. Bernsmann, F.; Ball, V.; Addiego, F.; Ponche, A.; Michel, M.; Gracio, J.J.D.A.; Toniazzo, V.; Ruch, D. Dopamine-melanin film deposition depends on the used oxidant and buffer solution. *Langmuir* **2011**, *27*, 2819–2825. [CrossRef] [PubMed]
26. Ponzio, F.; Barthes, J.; Bour, J.; Michel, M.; Bertani, P.; Hemmerle, J.; d'Ischia, M.; Ball, V. Oxidant control of polydopamine surface chemistry in acids: A mechanism-based entry to superhydrophilic-superoleophobic coatings. *Chem. Mater.* **2016**, *28*, 4697–4705. [CrossRef]
27. Zhang, L.; Mohammed, E.A.A.; Adriaens, A. Synthesis and electrochemical behavior of a magnesium fluoride-polydopamine-stearic acid composite coating on AZ31 magnesium alloy. *Surf. Coat. Technol.* **2016**, *307*, 56–64. [CrossRef]
28. Szaraniec, B.; Pielichowsk, K.; Pac, E.; Menaszek, E. Multifunctional polymer coatings for titanium implants. *Mater. Sci. Eng. C* **2018**, *93*, 950–957. [CrossRef] [PubMed]
29. Mayo, S.L.; Olafson, B.D.; Goddard, W.A. Dreiding: A generic force field for molecular simulations. *J. Phys. Chem.* **1990**, *94*, 8897–8909. [CrossRef]
30. Panteva, M.T.; Giambasu, G.M.; York, D.M. Force field for Mg^{2+}, Mn^{2+}, Zn^{2+}, and Cd^{2+} ions that have balanced interactions with nucleic acids. *J. Phys. Chem.* **2015**, *119*, 15460–15470. [CrossRef] [PubMed]
31. Kim, H.S.; Kim, G.H.; Kim, H.; Kim, W.J. Enhanced corrosion resistance of high strength Mg–3Al–1Zn alloy sheets with ultrafine grains in a phosphate-buffered saline solution. *Corros. Sci.* **2013**, *74*, 139–148. [CrossRef]
32. Pawar, S.; Zhou, X.; Thompson, G.E.; Scamans, G.; Fan, Z. The role of intermetallics on the corrosion initiation of twin roll cast AZ31 mg alloy. *J. Electrochem. Soc.* **2015**, *162*, C442–C448. [CrossRef]

33. Jiang, J.; Zhu, L.; Zhu, L.; Zhu, B.; Xu, Y. Surface characteristics of a self-polymerized dopamine coating deposited on hydrophobic polymer films. *Langmuir* **2011**, *27*, 14180–14187. [CrossRef] [PubMed]
34. Simons, W.W. *The Sadtler Handbook of Infrared Spectra*; Sadtler Research Laboratories: Philadelphia, PA, USA, 1978.
35. Orazem, M.E.; Orazem, M. *Electrochemical Impedance Spectroscopy*; John Wiley & Sons: Hoboken, NJ, USA, 2008.
36. Orishchin, N.; Crane, C.C.; Brownell, M.; Wang, T.; Jenkins, S.; Min Zou, A.N.C. Rapid deposition of uniform polydopamine coatings on nanoparticle surfaces with controllable thickness. *Langmuir* **2017**, *33*, 6046–6053. [CrossRef] [PubMed]
37. Ding, Y.H.; Floren, M.; Tan, W. Mussel-inspired polydopamine for bio-surface functionalization. *Biosurface Biotribol.* **2016**, *2*, 121–136. [CrossRef]
38. Kim, H.W.; McCloskey, B.D.; Choi, T.H.; Lee, C.; Kim, M.J.; Freeman, B.D.; Park, H.B. Oxygen concentration control of dopamine-induced high uniformity surface coating chemistry. *ACS Appl. Mater. Interfaces* **2013**, *5*, 223–238.
39. Lasia, A. *Electrochemical Impedance Spectroscopy and Its Applications*; Springer: New York, NY, USA, 2014.

© 2019 by the authors. Licensee MDPI, Basel, Switzerland. This article is an open access article distributed under the terms and conditions of the Creative Commons Attribution (CC BY) license (http://creativecommons.org/licenses/by/4.0/).

Article

Tri-Functional Calcium-Deficient Calcium Titanate Coating on Titanium Metal by Chemical and Heat Treatment

Seiji Yamaguchi *, Phuc Thi Minh Le, Morihiro Ito, Seine A. Shintani and Hiroaki Takadama

Department of Biomedical Sciences, College of Life and Health Sciences, Chubu University, Aichi 487-8501, Japan
* Correspondence: sy-esi@isc.chubu.ac.jp; Tel.: +81-568-51-6420

Received: 6 August 2019; Accepted: 27 August 2019; Published: 3 September 2019

Abstract: The main problem of orthopedic and dental titanium (Ti) implants has been poor bone-bonding to the metal. Various coatings to improve the bone-bonding, including the hydroxyapatite and titania, have been developed, and some of them have been to successfully applied clinical use. On the other hand, there are still challenges to provide antibacterial activity and promotion of bone growth on Ti. It was shown that a calcium-deficient calcium titanate coating on Ti and its alloys exhibits high bone-bonding owing to its apatite formation. In this study, Sr and Ag ions, known for their promotion of bone growth and antibacterial activity, were introduced into the calcium-deficient calcium titanate by a three-step aqueous solution treatment combined with heat. The treated metal formed apatite within 3 days in a simulated body fluid and exhibited antibacterial activity to *Escherichia coli* without showing any cytotoxicity in MC3T3-E1 preosteoblast cells. Furthermore, the metal slowly released 1.29 ppm of Sr ions. The Ti with calcium-deficient calcium titanate doped with Sr and Ag will be useful for orthopedic and dental implants, since it should bond to bone because of its apatite formation, promote bone growth due to Sr ion release, and prevent infection owing to its antibacterial activity.

Keywords: antibacterial activity; bone growth; apatite formation; titanium; silver; strontium; calcium titanate; ion release; cytotoxicity; controlled release

---

1. Introduction

Titanium metal (Ti) and its alloy are widely used for orthopedic and dental implants since they fulfill certain clinical needs from the point of view of mechanical properties, durability, and biocompatibility. Osseointegration occurs when the surfaces of the metals were roughened at micrometer scale such as 0.5–2.0, or 3.6–5.6 µm in calculated average roughness R_a, and 43–50 µm in maximum height R_z by plasma spraying, grid blasting and/or acid etching [1,2]. The nanometer-scale roughness produced by anodic oxidation also has been found to increase cell adhesion, proliferation and alkaline phosphatase activity [3,4]. Although these roughened Ti surfaces are able to directly contact with living bone, they still do not bond to it adequately. Mineralization process is a promising method to achieve strong and stable bone-bonding. It has been reported that bioactive glass/ceramics such as bioglass, hydroxyapatite, and glass ceramics A-W directly bonded to bone through the bone-like apatite layer formed on their surfaces [5].

Various types of surface coating of bioactive glass/ceramics by plasma spray, sputtering, sol-gel, and alternative soaking have been attempted [6–8]. Among them, a plasma spray coating of hydroxyapatite has been widely used to confer bone-bonding to total hip joint, dental implant, and so on. However, this does not form a stable bioactive surface layer, since the surfaces of the hydroxyapatite particles exposed to the plasma are partially melted, so the resultant calcium phosphate coating is liable to be decomposed in the living body in process of time [6].

Various types of surface modifications including alkali/acid solution and heat treatment, hydrothermal treatment, and ion implantation have been developed to confer apatite-forming capability on the metals so that the activated metals form bone-like apatite spontaneously on their surfaces by using the calcium and phosphate present in body fluids, and thereby bond to bone through the apatite [9–12]. Among the variety of the modification techniques, the alkali/acid solution and heat treatment has the needed characteristics for producing a uniform activated surface layer, even on the inner wall of a porous body, without requiring any especial apparatus [13,14]. It has been demonstrated that a bioactive sodium titanate layer is produced on Ti when the metal is soaked in 5 M NaOH solution at 60 °C for 24 h and subsequently heat-treated at 600 °C for 1 h [15,16]. The surface of treated metal forms a bone-like apatite spontaneously in the living body and bonds bone through this layer [16]. Total artificial hip joints (THAs) with the bioactive sodium titanate layer on their porous Ti layer have been under clinical use since 2007. A recent ten-year follow-up revealed the beneficial effects of the NaOH-heat-treated THAs to be a high survival rate (98%), no radiographic signs of loosening, and both early and stable bone-bonding [17]. However, two joints were retrieved owing to deep infection and periprosthetic femoral fracture, since the NaOH-heat-treated THAs neither promoted bone growth nor prevented infection [17]. Subsequently, the NaOH-heat treatment was modified to NaOH-CaCl$_2$-heat-water treatment to produce a calcium-deficient calcium titanate layer on Ti and Ti alloys, which resulted in more stable apatite formation and bone-bonding [18–20]. On the other hand, there are still the challenges of providing antibacterial activity and promoting bone growth on Ti. It has been reported that typically 1%–2% of patients with total hip arthroplasties suffer deep infections [21]. Dental peri-implant disease and infection have become a main focus of oral implantology [22].

Strontium (Sr) and silver (Ag) ion are candidates for the promotion of new bone formation and prevention of infection since the former exerts a therapeutic effect on osteoporosis bone due to increased new bone formation and deceased bone resorption, while the later prevents infection because of its strong effect against a broad range of microorganisms [23–26]. Pre-clinical study reports have shown that the Sr ions released from dosed strontium ranelate improve mineral density at various skeletal sites such as the total hip and lumbar spine, resulting in the improvement of osteoporosis [23,24]. The mechanism of the antimicrobial action of Ag ions is understood as resulting from an interaction with the thiol (sulfhydryl) groups in enzymes and proteins [25], and is effective even in the living body [26]. Studies have reported the separate incorporation of Sr or Ag into the surface of Ti and Ti alloys [27–30], but there are few reports of these ions being incorporated simultaneously. There are even reported attempts to incorporate these ions into the Ti surface in an effort to confer a capacity for apatite formation. It was reported that Ag-doped calcium phosphate coatings was produced on Ti by a combination of anodic oxidation, electrophoretic deposition, and magnetron-sputtering [31,32]. Although the coated metal exhibited strong antibacterial activity against *Escherichia coli* (*E. coli*), it a little decreased cell viability [32]. A novel method is desired to confer Ti the capabilities of excellent antibacterial activity without any decrease in cell viability, direct bone-bonding, and promotion of new bone formation at the same time.

In this study, Sr and Ag ions were introduced into the calcium-deficient calcium titanate produced on Ti under controlled conditions so that the treated Ti slowly released Sr and Ag ions in order to exhibit the functions of promoting new bone formation while preventing infection without decreasing apatite formation. The potential of the treated metal for clinical applications is discussed in terms of Sr and Ag ion release, antibacterial activity, cytocompatibility, and apatite formation.

2. Materials and Methods

2.1. Surface Treatment

Commercially pure Ti sections (Ti > 99.5%; Nilaco Co., Tokyo, Japan) 10 mm × 10 mm × 1 mm in size was grinded with #400 diamond plates and then cleaned in an ultrasonic bath by using acetone, 2-propanol and ultrapure water for 30 min, and dried at 40 °C overnight. They were immersed in

5 M NaOH (Reagent grade; Kanto Chemical Co., Inc., Tokyo, Japan) solution at 60 °C with shaking at 120 strokes/min for a period of 24 h followed by gentle rinse under ultrapure water flow for 30 s. The treated metals were soaked in a mixed solution consist of 50 mM $CaCl_2$ (Reagent grade; Kanto Chemical Co., Inc., Tokyo, Japan) and 50 mM $SrCl_2$ (Reagent grade; Kanto Chemical Co., Inc., Tokyo, Japan) at 40 °C, shaken at 120 strokes/min for 24 h, then washed and dried in a similar manner (designated as "Ca + Sr"). They were subsequently heated at 600 °C with programming rate of 5 °C/min and holding time of 1 h, then naturally cooled in an electric furnace. After the heat treatment, they were immersed in a mixed solution of 1 M $Sr(NO_3)_2$ (Reagent grade; Kanto Chemical Co., Inc., Tokyo, Japan) and X mM $AgNO_3$ (Reagent grade; Kanto Chemical Co., Inc., Tokyo, Japan) with an adjusted pH from 3 to 8 by adding HNO_3 or NH_3(aq) at 80 °C, where X is a range from 1 to 100 mM and designated as "Sr + X mM Ag, shaken , washed, and dried in the in a similar manner. The nominal and measured pH of the 1 M $Sr(NO_3)_2$ + 1 mM $AgNO_3$ are summarized in Table 1. Some of the Ti samples subjected to the same NaOH-Ca + Sr-heat treatment were subsequently soaked in 1 M $SrCl_2$ solution without Ag for comparison.

Table 1. Measured pH of 1 M $SrNO_3$ + 1 mM $AgNO_3$ solution corresponding to nominal pH.

Nominal pH	Measured pH	Additive
pH = 3	3.06	HNO_3
pH = 4	3.90	HNO_3
pH = 5	4.80	No additives
pH = 6	6.01	NH_3(aq)
pH = 7	7.16	NH_3(aq)
pH = 8	7.83	NH_3(aq)

2.2. Surface Analysis

2.2.1. Scanning Electron Microscopy and Energy Dispersive X-ray Analysis

The samples treated with the aqueous solution and heat were examined by field emission scanning electron microscopy (FE-SEM: S-4300, Hitachi Co., Tokyo, Japan) equipped with an energy dispersive X-ray spectrometer (EDX: EMAX-7000, Horiba Ltd., Kyoto, Japan). In FE-SEM observation, the samples were subjected to a thin-film coating of Pt–Pd, and 15 kV accelerate voltage was selected. In EDX analyisis, the elements of Ca, Ag, O, and Ti were quantified using 9 kV. The measurement was repeated on five different points, and their averaged value was used.

2.2.2. Thin-Film X-ray Diffraction and Fourier Transform Confocal Laser Raman Spectrometry

The surface structure of the samples subjected to the aqueous solution and heat treatment were analyzed by a thin-film X-ray diffractometer (TF-XRD: model RNT-2500, Rigaku Co., Tokyo, Japan) and Fourier transform confocal laser Raman spectrometer (FT-Raman: LabRAM HR800, Horiba Jobin Yvon, Longjumeau, France). In TF-XRD, the measurement was conducted at a power of 50 kV and 200 mA. A CuK was used as X-ray source and the incident beam angle was set to 1° against the sample surface. In FT-Raman, measurement was conducted with 514.5 nm Ar laser at 16 mW of power excitation.

2.2.3. Scratch Resistance

The scratch resistance of the surface layer to the metal substrate was examined by a thin-film scratch tester (CSR-2000, Rhesca Co., Ltd., Tokyo, Japan) according to JIS R-3255. A stylus with diameter of 5 μm and spring constant of 200 g/mm was pressed into the treated metal surface under the conditions of scratch speed of 10 μm/s, loading rates of 100 mN/min, and amplitude of 100 μm. Five measurements were performed on each sample, and their averaged values were used for analysis.

2.2.4. X-ray Photoelectron Spectroscopy

The allocation of elements such as Ag, Sr, Ca, Ti, O and C on the treated samples was analyzed using X-ray photoelectron spectroscopy (XPS, PHI 5000 Versaprobe II, ULVAC-PHI, Inc., Kanagawa, Japan) with Ar sputtering (spattering rate was 4 nm/min as SiO_2 conversion). In the analysis, the X-ray source of an Al-K radiation line was used with the take-off angle at 45°.

The obtained spectra were calibrated by 284.8 eV in binding energy of C 1s peak that is of the surfactant CH_2 groups on the substrate.

2.3. Ion Release

The treated samples were immersed in 2 mL of fetal bovine serum (FBS) (Gibco, Thermo Fisher Scientific, Waltham, MA, USA) with gently shaken at a speed of 50 strokes/min at 36.5 °C. After predetermined periods up to 14 days, the Sr^{2+} and Ag^+ ion concentrations in the FBS were determined by inductively coupled plasma emission spectroscopy (ICP, SPS3100, Seiko Instruments Inc., Chiba, Japan). The measurement was repeated 3 times for independently prepared samples, and their averaged values were calculated.

2.4. Soaking in Simulated Body Fluid (SBF)

After the aqueous solution and heat treatment the samples were immersed in 24 mL of acellular simulated body fluid (SBF) [33] that had been prepared according to ISO 23317. NaCl, $NaHCO_3$, KCl, $K_2HPO_4 \cdot 3H_2O$, $MgCl_2 \cdot 6H_2O$, $CaCl_2$ and Na_2SO_4 were purchased from Nacalai Tesque Inc., Kyoto, Japan and all of them were reagent grade. They were dissolved in fresh ultrapure water in this order, and their pH was adjusted exactly to 7.40 using tris(hydroxymethyl)aminomethane $(CH_2OH)_3CNH_2$ and 1 M HCl at 36.5 °C. After immersion periods of 3 days, the samples were rinsed and dried. Apatite formation formed on the metal surface was examined using FE-SEM, TF-XRD, and EDX.

2.5. Antibacterial Activity Test

The antibacterial activity of the treated Ti samples was evaluated by the film contact method (ISO22196) [34]. An E. coli (IFO 3972) suspension of 100 µL was inoculated on the treated Ti samples at 25 mm × 25 mm × 1 mm and then covered with a 20 mm × 20 mm polypropylene film that had been sterilized with ethanol and dried for 7 days in a clean bench. They were placed in a 100 mm diameter petri dish with a sterilized plastic cap filled with sterilized pure water to prevent drying of the bacterial suspension, and then stored in an incubator under 95% relative humidity at 35 °C for 24 h. After incubation, each sample was washed with 10 mL of a soybean casein digest broth containing lecithin and polyoxyethylene sorbitan monooleate (SCDLP broth) to collect the bacteria. The recovered suspension was subjected to ten-fold serial dilutions, followed by placed in petri dishes containing standard plate count agar at 35 °C for 48 h. After incubation, the number of viable E. coli was calculated using the dilution factor and the number of colonies that was counted on the petri dish. Finally, the antibacterial activity value (R) was calculated for each specimen as follows:

$$R = \{\log(B/A) - \log(C/A)\} = \log(B/C) \quad (1)$$

where A and B are the numbers of viable E. coli recovered from the untreated specimen immediately or 24 h incubation after inoculation. C is the number of viable E. coli recovered from the treated specimen immediately after 24 h incubation.

2.6. Cell Proliferation

MC3T3-E1 cells (subclone 14, ATCC, Manassas, VA, USA) were seeded onto Ti disk specimens that were 18 mm in diameter in 12-well plates at a density of 2×10^4 cells/well. They were cultured in α-MEM (Gibco, Thermo Fisher Scientific, Waltham, MA, USA) with 10% FBS and 1% penicillin/streptomycin at

37 °C in 5% CO_2 atmosphere. After 1 and 3 days, the cell count reagent SF (Nacalai tesque, Kyoto, Japan) was added to the medium and stored in the incubator for 2 h. After the incubation, 100 µL of the medium was taken to a 96-well plate. The absorbance at 450 nm that is attributed to formazan product derived from living cells was quantified by a Microplate reader (iMark™, Bio-Rad, Hercules, CA, USA). Four disk specimens were prepared for each sample type in this measurement.

2.7. Statistical Analysis

The obtained data in Section 2.6 was statistically analyzed by out using R language with these libraries (mvtnorm, survival, MASS, TH. data, multcomp, abind). The sample group data were initially tested for normality (Kolomogorov-Smirnov test) and homoscedasticity of variance (Bartlett's test). One-way analysis of variance (ANOVA) was adopted in the groups that satisfy those conditions to find any significant differences in the measured variables between control and treatment groups. When a difference was detected (p-value < 0.05), Tukey's multiple comparison test was performed to identify which treatment groups were significantly different. In this case, the ANOVA was satisfied in all analyses.

3. Results

3.1. Effect of the pH of the Solution Used on Apatite Formation

The chemical composition of the Ti surface after each aqueous solution and heat treatment was analyzed by EDX analysis. As shown in Table 2, 5.1% Na was induced by the initial NaOH treatment, and then replaced with 2.2% Ca and 1.3% Sr by the subsequent Ca + Sr treatment, which remained after the heat treatment. When the treated metal was immersed in 1 M $SrCl_2$ solution, amount of Sr a little increased, probably due to the additional induce of Sr into the surface because of high concentration of Sr ions in the solution. When the treated Ti specimens were soaked in 1 M $Sr(NO_3)_2$ and 1 mM $AgNO_3$ with a pH equal to or less than 4, 0.2% of the Ag was introduced into the surface, while the Ca amount was slightly decreased. In contrast, no decrease in the Ca content was observed when the Ti specimens were soaked in a solution with a pH greater than 4. The amount of Ag introuduced into the Ti surface tended to decrease with an increase in the pH of the solution.

Table 2. The results of EDX analysis on the surface of Ti subjected to Sr + 1 mM Ag treatment with various pH following NaOH, Ca + Sr and heat treatment.

Treatment	Element/at.%					
	O	Ti	Na	Ca	Sr	Ag
NaOH	65.1	29.8	5.1	0	0	0
NaOH-Ca + Sr	68.1	28.4	0	2.2	1.3	0
NaOH-Ca + Sr-heat	68.9	27.6	0	2.3	1.3	0
NaOH-Ca + Sr-heat-Sr + 1 mM Ag (pH = 3)	66.8	29.8	0	1.8	1.4	0.2
NaOH-Ca + Sr-heat-Sr + 1 mM Ag (pH = 4)	65.8	30.5	0	1.9	1.6	0.2
NaOH-Ca + Sr-heat-Sr + 1 mM Ag (pH = 5)	66.1	30.0	0	2.2	1.6	0.2
NaOH-Ca + Sr-heat-Sr + 1 mM Ag (pH = 6)	66.3	29.8	0	2.1	1.6	0.2
NaOH-Ca + Sr-heat-Sr + 1 mM Ag (pH = 7)	65.6	30.6	0	2.1	1.6	0.1
NaOH-Ca + Sr-heat-Sr+1 mM Ag (pH = 8)	65.7	30.4	0	2.1	1.7	0.1
NaOH-Ca + Sr-heat-Sr	68.7	27.4	0	2.2	1.7	0

The standard deviation of each element is as follows (SD_i: i indicates individual element). SD_O < 0.44, SD_{Ti} < 0.44, SD_{Ca} < 0.12, SD_{Sr} < 0.11, SD_{Ag} < 0.08.

The surface structure of these samples was examined by XRD analysis and Raman scattering as shown in Figure 1. Sodium hydrogen titanate (SHT; $Na_xH_{2-x}Ti_3O_7$) was produced after the initial NaOH treatment. These XRD and Raman profiles were not apparently changed except for a slight shift of about 920 to 900 cm^{-1} in Raman by the subsequent Ca + Sr treatment. Since the Raman peak around 920 cm^{-1} in SHT was attributed to Ti–O bonds coordinated with Na ions [35], the results indicate that the SHT transformed into Sr-containing calcium hydrogen titanate by replacing Na with Ca and Sr without any apparent change of its structural frame. This material was dehydrated by the subsequent heat treatment to form Sr-containing calcium titanate and rutile accompanied by a small quantity of anatase. Although the XRD and Raman profiles were apparently unchanged by the final Sr + 1 mM Ag treatment regardless of the pH of the solution, it may be inferred that the Sr-containing calcium titanate transformed into Sr- and Ag-containing calcium titanate or Sr- and Ag-containing calcium-deficient calcium titanate by a final solution treatment with a pH ≥ 5 or pH ≤ 4, respectively, according to surface chemical composition, as shown in Table 2. When these samples were subjected to scratch resistance test, the surface layer formed after the first NaOH treatment showed low scratch resistance value as 0.9 ± 0.5 mN. This value was almost unchanged by the second solution treatment (the value was 1.6 ± 0.5 mN). In contrast, it markedly increased to 37.8 ± 7.0 mN after heat and remained even after the Sr + 1 mM Ag (pH = 4) treatment (the value was 39.3 ± 3.7 mN).

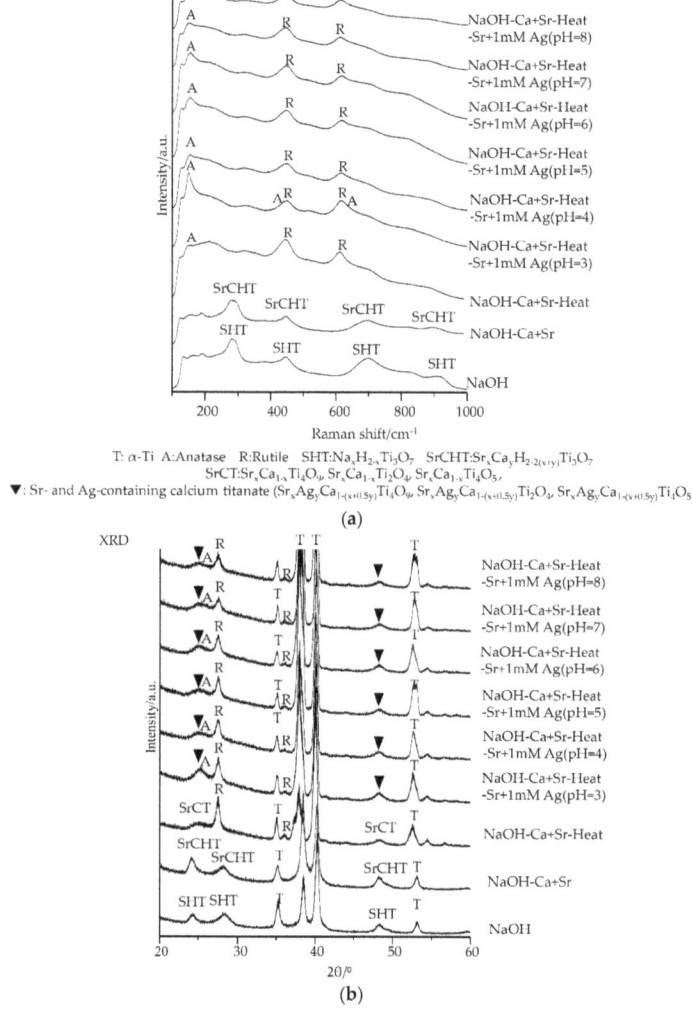

Figure 1. Raman (**a**) and XRD (**b**) spectra of Ti surfaces subjected to Sr + 1 mM Ag treatment with various pH following NaOH, Ca + Sr and heat treatment.

Depth profile of XPS analysis on the metal sample after the NaOH-Ca + Sr-heat-Sr + 1 mM Ag (pH = 4) is shown in Figure 2. Comparable amounts of Sr and Ca and a small amount of Ag were detected near the top surface and decreased gradually in depth until approximately 1 μm. The results are consistent with the surface chemical composition in Table 2, and the thickness of the surface layer on cross sectional SEM observation, where an approximately 1 μm thick surface layer was evident (data not shown). Figure 3 shows narrow XPS spectra of the treated metal. The peaks at 367.7 and 373.8 eV attributed to Ag_2O [36] were observed, verifying that Ag was incorporated into the surface as a form of Ag^+ ion.

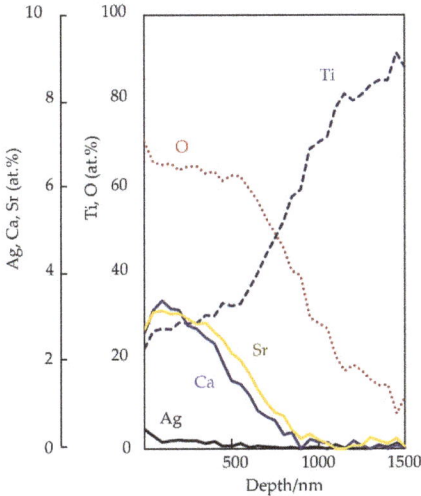

Figure 2. XPS depth profile of the surface of Ti subjected to NaOH-Ca + Sr-heat-Sr + 1 mM Ag (pH = 4).

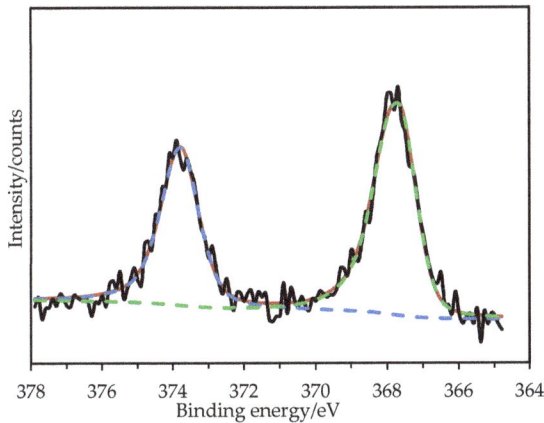

Figure 3. Narrow XPS Ag 3d profile of Ti surface subjected to NaOH-Ca + Sr-heat-Sr + 1 mM Ag (pH = 4). Red solid line: composite line of blue and green dot lines, Blue dot line: deconvolution line of Ag $3d_{3/2}$, Green dot line: deconvolution line of Ag $3d_{5/2}$.

Figure 4 shows the SEM images of the Ti surface before and after soaking in SBF for 3 days that was subjected to Sr + 1 mM Ag treatment with various pH levels following NaOH, $CaCl_2$, and heat treatment. It can be seen that a similar network morphology on a nanometer scale was produced by the aqueous solution and heat treatment regardless of the pH of the 1 M $Sr(NO_3)_2$ + 1 mM $AgNO_3$

solution used in the final solution treatment. When the treated metals were immersed in SBF for 3 days, apatite formation was observed only on the surfaces that had been treated with 1 M Sr(NO$_3$)$_2$ and 1 mM AgNO$_3$ with a pH equal to or less than 4. In terms of apatite formation as well as the Ag and Sr content, the pH of the aqueous solution was fixed at 4 in the following experiment.

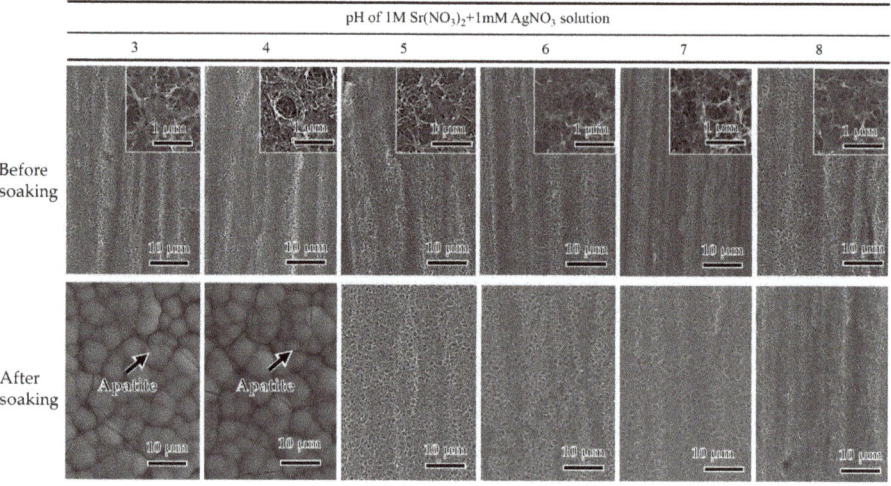

Figure 4. SEM images of the surfaces of Ti before and after soaking in SBF for 3 days that has been subjected to Sr + 1 mM Ag treatment with various pH following NaOH, Ca + Sr and heat treatment. Small windows show high magnification images on Ti before soaking in SBF. The digits in the table above the images stand for pH of 1 M Sr(NO$_3$)$_2$ + 1 mM AgNO$_3$ solution in the final solution treatment.

3.2. Effect of the Ag Concentration in the Solution Treatment on Apatite Formation

The Ti samples subjected to the NaOH-Ca + Sr-heat treatment were soaked in a 1 M Sr(NO$_3$)$_2$ solution of pH = 4 with different concentrations of AgNO$_3$ from 1 to 100 mM added, and their surface chemical composition was analyzed by EDX, as shown in Table 3. The amount of Ag increased, with increasing concentration of Ag in the final solution treatment up to 1.1% along with slightly decreased Sr content.

Table 3. The results of EDX analysis on the surface of Ti subjected to Sr + X mM Ag (pH = 4) treatment (X = 1–100) following NaOH, Ca + Sr and heat treatment.

Treatment	Element/at.%				
	O	Ti	Ca	Sr	Ag
NaOH-Ca + Sr-heat-Sr + 1 mM Ag (pH = 4)	65.8	30.5	1.9	1.6	0.2
NaOH-Ca + Sr-heat-Sr + 2 mM Ag (pH = 4)	66.1	30.3	1.9	1.5	0.2
NaOH-Ca + Sr-heat-Sr + 5 mM Ag (pH = 4)	66.1	30.3	1.9	1.5	0.3
NaOH-Ca + Sr-heat-Sr + 10 mM Ag (pH = 4)	66.3	30.2	1.8	1.4	0.3
NaOH-Ca + Sr-heat-Sr + 20 mM Ag (pH = 4)	66.1	30.2	1.9	1.4	0.4
NaOH-Ca + Sr-heat-Sr + 50 mM Ag (pH = 4)	66.1	29.6	2.0	1.5	0.9
NaOH-Ca + Sr-heat-Sr + 100 mM Ag (pH = 4)	65.7	29.9	2.0	1.3	1.1

The standard deviation of each element is as follows (SD$_i$: i indicates individual element). SD$_O$ < 0.6, SD$_{Ti}$ < 0.4, SD$_{Ca}$ < 0.1, SD$_{Sr}$ < 0.1, SD$_{Ag}$ < 0.1.

SEM revealed that nano sized particles started to be precipitated on the surface of Ti when the Ag concentration in the final solution treatment was 20 mM, and their number increased with an increasing Ag concentration, as shown in Figure 5. These particles were determined to be metallic Ag particles by EDX line analysis (data not shown). It can be seen from the XRD and Raman spectra of the treated samples (Figure 6) that a peak at around 44° attributed to metallic Ag [37] was detected on the Ti surface only when the Ag concentration in the final solution treatment was equal or greater than 20 mM. There were no other changes that depended on the Ag concentration of the final solution treatment in the XRD and Raman profiles.

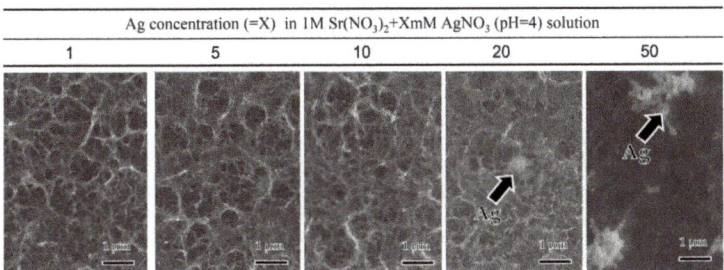

Figure 5. SEM images of the surfaces of Ti subjected to Sr + X mM Ag (pH = 4) treatment (X = 1–50) following NaOH, Ca + Sr and heat treatment.

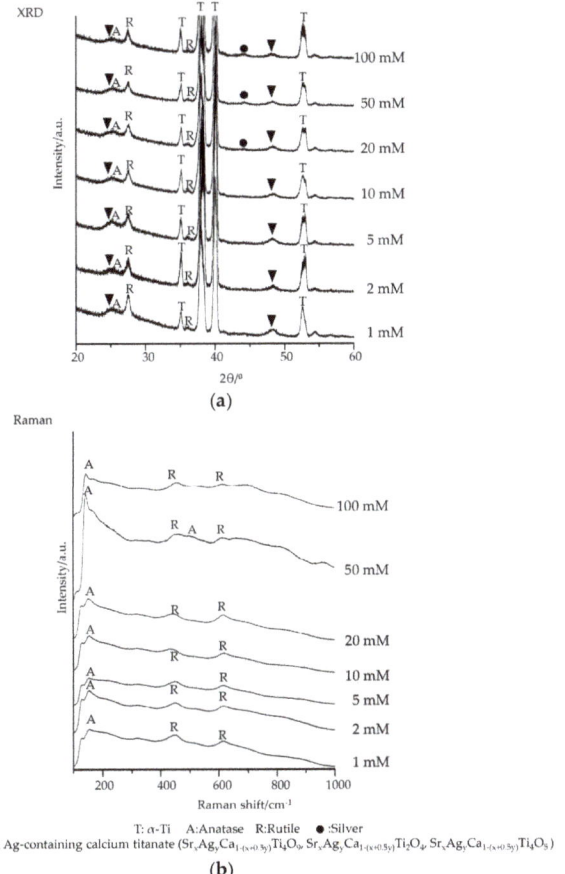

Figure 6. XRD (a) and Raman (b) spectra of Ti surfaces subjected to Sr + X mM Ag (pH = 4) treatment with various Ag concentrations (X = 1–100) following NaOH, Ca + Sr and heat treatment.

When these samples were soaked in SBF, they formed spherical particles on their surfaces within 3 days that were identified as low crystalline apatite by XRD (data not shown), regardless of the Ag content and even in the presence of Ag particles as shown in Figure 7.

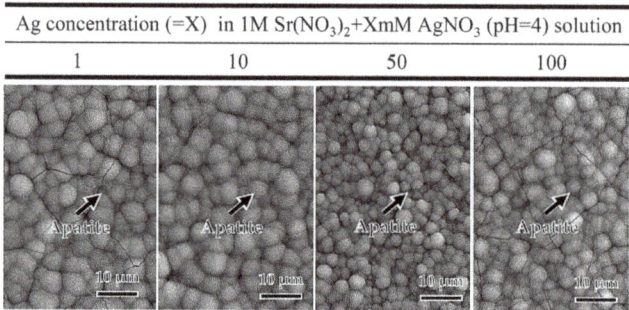

Figure 7. SEM images of the surface of Ti before and after soaked in SBF for 3 days that has been subjected to Sr + X mM Ag (pH = 4) treatment with various Ag concentrations (X = 1–50) following NaOH, CaCl$_2$, and heat treatment.

3.3. Effect of the Ag Concentration in the Solution on Cytotoxicity

The Ti samples with Sr- and Ag-containing calcium-deficient calcium titanate without any metallic Ag particles were prepared by Sr + 1 mM Ag (pH = 4) or Sr + 10 mM Ag (pH = 4) treatment following NaOH-Ca + Sr-heat treatment, and their effect on the viability of MC3T3-E1 cells was examined. The results were compared with those on untreated or NaOH-Ca + Sr-heat-Sr-treated Ti with Sr-containing calcium titanate free of Ag. As shown in Figure 8, the cell viability significantly increased in the treated Ti subjected to NaOH-Ca + Sr-heat-Sr + 1 mM Ag (pH = 4) compared with untreated samples in the culture period of 1 day. There were no significant differences between the treated samples. At 3 days, although all of the treated samples showed higher cell viability than the untreated sample, there was a difference between the treated samples: NaOH-Ca + Sr-heat-Sr was highest, followed by NaOH-Ca + Sr-heat-Sr + 1 mM Ag (pH = 4) and then NaOH-Ca + Sr-heat-Sr + 10 mM Ag (pH = 4). There a significant difference between Ti samples subjected to the NaOH-Ca + Sr-heat-Sr and NaOH-Ca + Sr-heat-Sr + 10 mM Ag (pH = 4) treatments.

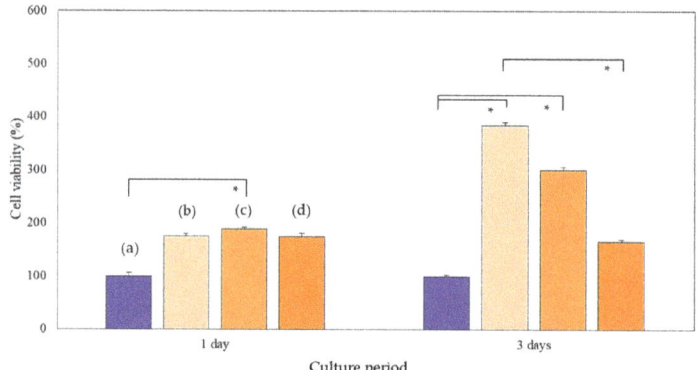

Figure 8. Cell viability of MC3T3-E1 on Ti (a) untreated and subjected to (b) NaOH-CaCl$_2$-heat-Sr, (c) NaOH-CaCl$_2$-heat-Sr + 1 mMAg (pH = 4), and (d) NaOH-CaCl$_2$-heat-Sr + 10 mM Ag (pH = 4). Asterisk stands for statistically significant difference ($p < 0.05$).

3.4. Antibacterial Activity

The antibacterial activity against *E. coli* of the Ti subjected to NaOH-Ca + Sr-heat-Sr + 1 mM Ag (pH = 4) was examined by the film contact method. As a result, the treated Ti displayed a 5.9-log reduction compared with the untreated Ti, as shown in Table 4, indicating sufficiently high antibacterial activity.

Table 4. Antibacterial activity results on Ti untreated and subjected to NaOH-Ca + Sr-heat-Sr + 1 mM Ag (pH = 4).

Treatment	Average of *E. Coli* count/CFU		Antibacterial Activity Value
	After Inoculation	After Incubation	
Untreated	2.8×10^6	1.5×10^7	–
NaOH-Ca + Sr-heat-Sr + 1 mM Ag (pH = 4)	4.7×10^6	<20	5.9

3.5. Ion Release Test

The same treated samples were soaked in FBS for up to 14 days and the Sr and Ag ions released from the samples were measured by ICP. It can be seen in Figure 9 that the treated metal released 0.78 ppm of Ag and 0.87 ppm of Sr within 1 h, and then slowly released another 0.91 ppm of Ag and 0.42 ppm of Sr over 14 days.

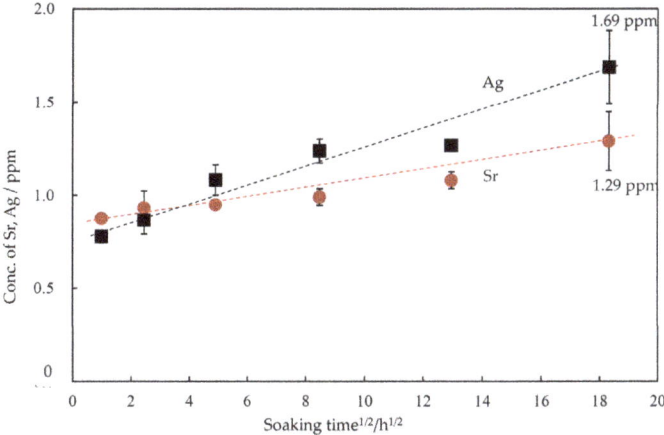

Figure 9. Concentrations of Sr and Ag ions released from Ti subjected NaOH-CaCl$_2$-heat-Sr + 1 mM Ag (pH = 4), as a function of square root of soaking time in FBS, which was measured by ICP. Black squares and red circles indicate Ag and Sr, respectively.

4. Discussion

The SHT (Na$_x$H$_{2-x}$Ti$_3$O$_7$) formed on Ti and its alloys by NaOH treatment has potent ion exchange capacity because of its layered structure [38]. It has been reported that the Na$^+$ in SHT can be exchanged by various types of and even various valences of metal ions such as Ag$^+$, magnesium (Mg^{2+}), Ca^{2+}, Sr^{2+}, gallium (Ga^{3+}), and more than two types of them simultaneously [27–30,39–42]. The present study proved that even three types of functional ions such as Ca, Sr and Ag can be controllably incorporated into the surface of Ti by a combination of simple aqueous solution and heat treatment.

The incorporation of Ag$^+$ ions into SHT was first attempted by Inoue et al. [43]. They formed an Ag-containing sodium titanate layer by soaking Ti in 0.05 M silver acetate solution following NaOH solution or NaOH hydrothermal treatment. The treated metal contained Ag$^+$ ions, but also a certain amount of precipitated metallic silver particles [43]. Such colloidal metallic particles are undesirable because they could be released and transferred to other organ, resulting in side effects.

Kizuki et al. [28] and Prabu et al. [44] demonstrated that Ag can be induced into SHT formed on Ti and Ti–6Al–4V alloy as ion form by soaking the metals in 0.01–100 mM Ag(NO)$_3$ solution after the NaOH solution. However, metallic Ag particles were formed again when the treated metals were subsequently heat-treated at 600 °C [28]. Eventually, Ag-containing calcium titanate free of Ag particles forms on Ti and Ti–15Zr–4Nb–4Ta alloy as the result of immersing in 1 mM AgNO$_3$ solution following NaOH-CaCl$_2$-heat treatment [28]. The treated Ti and Ti–15Zr–4Nb–4Ta contained 0.55% and 0.27% of Ag on their surfaces and released 2.66 and 1.45 ppm of Ag$^+$ ions into FBS, respectively. These metals showed strong antibacterial activity against *S. aureus* (more than 99% reduction). However, the cytotoxicity of these treated metals was not reported.

In the present study, Ti along with Sr- and Ag-containing calcium titanate with various amounts of Ag were produced. It is shown in Table 3 and Figures 5 and 6 that the Ag content increased from 0.2% to 1.1% in the final solution treatment, while metallic Ag precipitated when the Ag content on the surface of Ti became 0.4%. Thus, Ti specimens with the surface containing 0.2% and 0.3% Ag in the ion form were prepared and their cytotoxicity compared with the untreated and the Ag free Ti with the Sr-containing calcium titanate. As a result, the Ti with the Sr-containing calcium titanate was shown to markedly increase the cell viability after 3 days compared with the untreated Ti. This is consistent with our previous report in which the cell viability in the Ti subjected to the same treatment was significantly increased compared with untreated Ti on 5 days [45]. A similar increase in cell viability was observed in the case of Ti with 0.2% Ag at 1 and 3 days. In contrast, the cell viability of Ti with 0.3% Ag was significantly lower than that of Ti with Sr-containing calcium titanate, although it was comparable to that of untreated Ti. This is probably because the Ag$^+$ ions that were released from the surface of the treated Ti accumulated in the vicinity of the surface and suppressed cell proliferation. The Ti with the 0.2% Ag resulting from NaOH-Ca + Sr-heat-Sr + 1 mM Ag (pH = 4) slowly released 1.69 ppm Ag$^+$ ions into FBS over 14 days and exhibited potent antibacterial activity against *E. coli*, as shown in Figure 9 and Table 4. It has been reported that Ti and its alloys with 0.27%–0.67% Ag content induced by chemical and heat treatment exhibited strong antibacterial activity without any cytotoxicity [46]. The results in this study are consistent with these reports. On the other hand, they also imply that a lower Ag content, such as 0.3%, may suppress cell proliferation without any cytotoxicity.

Bone-bonding is a crucial function of an implant and may be predicted by examining the apatite formation on the material that occurs in SBF [33]. Fujibayashi et al. reported that bioactive Na$_2$O-CaO-SiO$_2$ glass powders with different compositions and induction periods of apatite formation in SBF induced different amount of new bone formation: amount of new bone formation increased with decreasing induction periods of apatite formation in SBF [47]. They recommended the materials able to form apatite within 3 days in SBF for practical use. It should be noted that the bone-bonding strength might be affected by various factors including strength and thickness of the substrate and coating layers [48]. It was reported that Ti–15Zr–4Nb–4Ta alloy with approximately 0.5 μm calcium-deficient calcium titanate layer implanted into rabbit tibia showed lower critical detaching load in detaching test than Ti with approximately 1 μm calcium-deficient calcium titanate layer at 4 weeks of implantation period, although both of them exhibited direct bone-bonding in histological observation [19,20]. The critical detaching load increased with increasing implantation periods up to 26 weeks in both cases, where fracture occurred not at interfaces between the treated metals and bone but inside the bone [20]. In this study, abundant apatite formation was observed on the treated Ti within 3 days in SBF regardless of the Ag content and even in the presence of metallic Ag particles. Sufficiently high bone-bonding is expected on these metals. On the other hand, apatite formation strongly depended on the pH of the solution in the final solution treatment: Ti formed apatite only when it had been soaked in 1 M Sr(NO$_3$)$_2$ and 1 mM AgNO$_3$ with a pH equal to or less than 4, as shown in Figure 7. This might be due to the formation of Sr- and Ag-containing calcium-deficient calcium titanate on Ti. It is reported that the calcium-deficient calcium titanate that forms on Ti exhibits an increased capacity for apatite formation compared with calcium titanate because of its greater release of Ca^{2+} ions [28,41].

It is expected that new bone growth surrounding Ti will be accelerated if appropriate concentrations of Sr^{2+} ions are released from Ti in the living body. The new bone tightly bonds to the metal via the apatite that have been formed on the metal surface. In the present study, the Ti subjected to NaOH-Ca + Sr-heat-Sr + 1 mM Ag (pH = 4) slowly released 1.29 ppm of Sr^{2+} ions into FBS in addition to Ag^+ ions over 14 days. This value falls in the effective range of 0.21 and 21.07 ppm that was shown to enhance the expression of a key osteoblast transcription factor gene (Cbfa1) and alkaline phosphatase (ALP) activity in human bone marrow mesenchymal stem cells [49]. Park et al. hydrothermally produced $SrTiO_3$ coating on Ti that released 0.75 ppm of Sr^{2+} ions into physiological saline solution. When primary mouse bone marrow stromal cells were cultured on the metal surface, increased cell activity including cell attachment, spreading, gene expression, and ALP activity was shown [30]. Yamaguchi et al. [29] and Okuzu et al. [45] reported that Ti having Sr-containing calcium titanate on its surface increased proliferation and osteogenic differentiation of MC3T3-E1 cells by releasing 0.92 ppm of Sr^{2+} ions. In their reports, various types of gene expression, including integrin β1, β catenin, cyclin D1 and ALP were up-regulated and resulted in extracellular mineralization. They also showed that biomechanical strength as well as bone-implant contact became greater than the Ti with calcium-deficient calcium titanate, when the metals were implanted into rabbit tibia at short periods of 4–8 weeks.

Based on these results, the Ti with Sr- and Ag-containing calcium-deficient calcium titanate is expected to form apatite on its surface and bond to living bone through the apatite, while promoting new bone growth by releasing Sr^{2+} ions. Furthermore, it should prevent postoperative infection because of its antibacterial activity.

5. Conclusions

Tri-functional bioactive Ti with the Sr- and Ag-containing calcium-deficient calcium titanate was produced by a combination of aqueous solution and heat treatment. An effective amount of Ca, Sr and Ag was introduced into the surface of Ti by controlling the ion concentration and pH of the solution so that the treated Ti precipitated apatite in SBF within 3 days and exhibited strong antibacterial activity, with increased cell viability. Furthermore, it released Sr^{2+} ions into FBS at a level up to 1.29 ppm. This type of multifunctional Ti is promising for the next generation of orthopedic and dental implants in next generation.

Author Contributions: Conceptualization, S.Y., M.I., S.A.S. and H.T.; Methodology, S.Y., P.T.M.L. and M.I.; Software, S.Y. and S.A.S.; Validation, S.Y., P.T.M.L., M.I. and S.A.S.; Formal Analysis, S.Y., M.I. and S.A.S.; Investigation, S.Y., P.T.M.L. and M.I.; Resources, S.Y. and M.I.; Data Curation, S.Y. and S.A.S.; Writing—Original Draft Preparation, S.Y.; Writing—Review and Editing, S.Y.; Visualization, S.Y.; Supervision, M.I. and H.T.; Project Administration, S.Y.; Funding Acquisition, S.Y.

Funding: This research was partially supported by Chubu University Grant (B) 19M02B.

Conflicts of Interest: The authors declare no conflict of interest.

References

1. Hacking, S.A.; Tanzer, M.; Harvey, E.J.; Krygier, J.J.; Bobyn, J.D. Relative contributions of chemistry and topography to the osseointegration of hydroxyapatite coatings. *Clin. Orthopaed. Relat. Res.* **2002**, *405*, 24–38. [CrossRef] [PubMed]
2. Coelho, P.G.; Granjeiro, J.M.; Romanos, G.E.; Suzuki, M.; Silva, N.R.F.; Cardaropoli, G.; Thompson, V.P.; Lemons, J.E. Basic research methods and current trends of dental implant surfaces. *J. Biomed. Mater. Res.* **2009**, *88*, 579–596. [CrossRef] [PubMed]
3. Brammera, K.S.; Ohd, S.; Cobba, C.J.; Bjurstenb, L.M.; van der Heydec, H.; Jina, S. Improved bone-forming functionality on diameter-controlled TiO_2 nanotube surface. *Acta Biomater.* **2009**, *5*, 3215–3223. [CrossRef]
4. Zhao, L.; Mei, S.; Chu, P.K.; Zhang, Y.; Wu, Z. The influence of hierarchical hybrid micro/nano-textured titanium surface wih titania nanotubes on osteoblast functions. *Biomaterials* **2010**, *31*, 5072–5082. [CrossRef] [PubMed]
5. Kokubo, T. *Bioceramics and Their Clinical Applications*; Woodhead Publishing: Cambridge, UK, 2008.

6. Kawai, T.; Takemoto, M.; Fujibayashi, S.; Tanaka, M.; Akiyama, H.; Nakamura, T.; Matsuda, S. Comparison between alkali heat treatment and sprayed hsydroxyapatite coating on thermally-sprayed rough Ti surface in rabbit model: Effects on bone-bonding ability and osteoconductivity. *J. Biomed. Mater. Res. Part B Appl. Biomater.* **2015**, *103*, 1069–1081. [CrossRef]
7. Leeuwenburgh, S.C.G.; Wolke, J.G.C.; Jansen, J.A.; de Groot, K. *Bioceramics and Their Clinical Applications*; Kokubo, T., Ed.; Woodhead Publishing: Cambridge, UK, 2008; Chapter 20; pp. 464–484.
8. Strange, D.G.T.; Oyen, M.L. Biomimetic bone-like composites fabricated through an automated alternate soaking process. *Acta Biomater.* **2011**, *7*, 3586–3594. [CrossRef] [PubMed]
9. Nayab, S.N.; Jones, F.H.; Olsen, I. Modulation of the human bone cell cycle by calcium ion-implantation of titanium. *Biomaterials* **2007**, *28*, 38–44. [CrossRef] [PubMed]
10. Tsutsumi, Y.; Niinomi, M.; Nakai, M.; Tsutsumi, H.; Doi, H.; Nomura, N.; Hanawa, T. Micro-arc oxidation treatment to improve the hard-tissue compatibility of Ti–29Nb–13Ta–4.6Zr alloy. *Appl. Surf. Sci.* **2012**, *262*, 34–38. [CrossRef]
11. Park, J.W.; Park, K.B.; Suh, J.Y. Effects of calcium ion incorporation on bone healing of Ti6Al4V alloy implants in rabbit tibiae. *Biomaterials* **2007**, *28*, 3306–3313. [CrossRef] [PubMed]
12. Park, J.W.; Kim, Y.J.; Jang, J.H.; Kwon, T.G.; Bae, Y.C.; Suh, J.Y. Effects of phosphoric acid treatment of titanium surfaces on surface properties, osteoblast response and removal of torque forces. *Acta Biomater.* **2010**, *6*, 1661–1670. [CrossRef] [PubMed]
13. Takemoto, M.; Fujibayashi, S.; Neo, M.; Suzuki, J.; Matsushita, T.; Kokubo, T.; Nakamura, T. Osteoinductive porous titanium implants: Effect of sodium removal by dilute HCl treatment. *Biomaterials* **2006**, *27*, 2682–2691. [CrossRef] [PubMed]
14. Kawai, T.; Takemoto, M.; Fujibayashi, S.; Akiyama, H.; Yamaguchi, S.; Pattanayak, D.K.; Doi, K.; Matsushita, T.; Nakamura, T.; Kokubo, T.; et al. Osteoconduction of porous Ti metal enhanced by acid and heat treatments. *J. Mater. Sci. Mater. Med.* **2013**, *24*, 1707–1715. [CrossRef] [PubMed]
15. Kokubo, T.; Miyaji, F.; Kim, H.M.; Nakamura, T. Spontaneous formation of bonelike apatite layer on chemically treated titanium metals. *J. Am. Ceram. Soc.* **1996**, *79*, 1127–1129. [CrossRef]
16. Kokubo, T.; Yamaguchi, S. Novel bioactive materials developed by simulated body fluid evaluation: Surface-modified Ti metal and its alloys. *Acta Biomater.* **2016**, *44*, 16–30. [CrossRef] [PubMed]
17. So, K.; Kaneuji, A.; Matsumoto, T.; Matsuda, S.; Akiyama, H. Is the bone-bonding ability of a cementless total hip prosthesis enhanced by alkaline and heat treatments? *Clin. Orthop. Relat. Res.* **2013**, *471*, 3847–3855. [CrossRef] [PubMed]
18. Kizuki, T.; Matsushita, T.; Kokubo, T. Preparation of bioactive Ti metal surface enriched with calcium ions by chemical treatment. *Acta Biomater.* **2010**, *6*, 2836–2842. [CrossRef]
19. Yamaguchi, S.; Takadama, H.; Matsushita, T.; Nakamura, T.; Kokubo, T. Apatite-forming ability of Ti–15Zr–4Nb–4Ta alloy induced by calcium solution treatment. *J. Mater. Sci. Mater. Med.* **2010**, *21*, 439–444. [CrossRef]
20. Fukuda, A.; Takemoto, M.; Saito, T.; Fujibayashi, S.; Neo, M.; Yamaguchi, S.; Kizuki, T.; Matsushita, T.; Niinomi, M.; Kokubo, T.; et al. Bone bonding bioactivity of Ti metal and Ti–Zr–Nb–Ta alloys with Ca ions incorporated on their surfaces by simple chemical and heat treatments. *Acta Biomater.* **2011**, *7*, 1379–1386. [CrossRef]
21. Hamilton, H.; Jamieson, J. Deep infection in total hip arthroplasty. *Can. J. Surg.* **2008**, *51*, 111–117.
22. O'Neill, E.; Awale, G.; Daneshmandi, L.; Umerah, O.; Lo, K.W. The roles of ions on bone regeneration. *Drug Discov. Today* **2018**, *23*, 879–890. [CrossRef]
23. Barbara, A.; Delannoy, P.; Denis, B.G.; Marie, P.J. Normal matrix mineralization induced by strontium ranelate in MC3T3-E1 osteogenic cells. *Metabolism* **2004**, *53*, 532–537. [CrossRef] [PubMed]
24. Bonnelye, E.; Chabadel, A.; Saltel, F.; Jurdic, P. Dual effect of strontium ranelate: Stimulation of osteoblast differentiation and inhibition of osteoclast formation and resorption in vitro. *Bone* **2008**, *42*, 129–138. [CrossRef] [PubMed]
25. Jung, W.K.; Koo, H.C.; Kim, K.W.; Shin, S.; Kim, S.H.; Park, Y.H. Antibacterial activity and mechanism of action of the silver ion in staphylococcus aureus and *Escherichia Coli*. *Appl. Envrion. Microbiol.* **2008**, *74*, 2171–2178. [CrossRef] [PubMed]
26. Klasen, H.J. Historical review of the use of silver in the treatment of burns. *Burns* **2000**, *26*, 117–130. [CrossRef]

27. Ferraris, S.; Venturello, A.; Miola, M.; Cochis, A.; Rimondini, L.; Spriano, S. Antibacterial and bioactive nanostructured titanium surfaces for bone integration. *Appl. Surf. Sci.* **2014**, *311*, 279–291. [CrossRef]
28. Kizuki, T.; Matsushita, T.; Kokubo, T. Antibacterial and bioactive calcium titanate layers formed on Ti metal and its alloys. *J. Mater. Sci. Mater. Med.* **2014**, *25*, 1737–1746. [CrossRef] [PubMed]
29. Yamaguchi, S.; Nath, S.; Matsushita, T.; Kokubo, T. Controlled release of strontium ions from a bioactive Ti metal with a Ca-enriched surface layer. *Acta Biomater.* **2014**, *10*, 2282–2289. [CrossRef] [PubMed]
30. Park, J.W.; Kim, Y.J.; Jang, J.H.; Suh, J.Y. Surface characteristics and primary bone marrow stromal cell response of a nanostructured strontium-containing oxide layer produced on a microrough titanium surface. *J. Biomed. Mater. Res. A* **2012**, *100*, 477–487. [CrossRef] [PubMed]
31. Chernozema, R.V.; Surmeneva, M.A.; Krauseb, B.; Baumbach, T.; Ignatov, V.P.; Prymak, O.; Loza, K.; Epple, M.; Ennen-Roth, F.; Wittmar, A.; et al. Functionalization of titania nanotubes with electrophoretically deposited silver and calcium phosphate nanoparticles: Structure, composition and antibacterial assay. *Mater. Sci. Eng. C* **2019**, *97*, 420–430. [CrossRef]
32. Surmeneva, M.A.; Sharonova, A.A.; Chernousova, S.; Prymak, O.; Loza, K.; Tkachev, M.S.; Shulepov, I.A.; Epple, M.; Surmenev, M.A. Incorporation of silver nanoparticles into magnetron-sputtered calcium phosphate layers on titanium as an antibacterial coating. *Colloids Surf. B Biointerfaces* **2017**, *156*, 104–113. [CrossRef]
33. Kokubo, T.; Takadama, H. How useful is SBF in predicting in vivo bone bioactivity? *Biomaterials* **2006**, *27*, 2907–2915. [CrossRef] [PubMed]
34. *ISO22196:2011 Measurement of Antibacterial Activity on Plastics and Other Non-Porous Surfaces*; ISO: Geneva, Switzerland, 2011.
35. Kolwn'ko, Y.V.; Kovnir, K.A.; Gavrilov, A.I.; Garshev, A.V.; Frantti, J.; Lebedev, O.I.; Churagulov, B.R.; Tendeloo, G.V.; Yoshimura, M. Hydrothermal synthesis and characterization of nanorods of various titanates and titanium dioxide. *J. Phys. Chem. B* **2006**, *110*, 4030–4038. [CrossRef]
36. *NIST X-ray Photoelectron Spectroscopy Database*, version 4.1; NIST: Gaithersburg, MA, USA, 2012.
37. *Powder Diffraction Data File 00-004-0783. Joint Committee on Powder Diffraction Standards (JCPDS)*; International Centre for Diffraction Data: Newtown Square, PA USA, 2015.
38. Kokubo, T.; Yamaguchi, S. Novel bioactive titanate layers formed on Ti metal and its alloy by chemical treatments. *Materials* **2010**, *3*, 48–63. [CrossRef]
39. Yamaguchi, S.; Matsushita, T.; Nakamura, T.; Kokubo, T. Bioactive Ti metal with Ca-enriched surface layer able to release Zn ion. *Key Eng. Mater.* **2013**, *529*, 547–552. [CrossRef]
40. Yamaguchi, S.; Nath, S.; Sugawara, Y.; Divakarla, K.; Das, T.; Manos, J.; Chrzanowski, W.; Matsushita, T.; Kokubo, T. Two-in-one biointerfaces—Antimicrobial and bioactive nanoporous gallium titanate layers for titanium implants. *Nanomaterials* **2017**, *7*, 229. [CrossRef] [PubMed]
41. Yamaguchi, S.; Matsushita, T.; Kokubo, T. A bioactive Ti metal with a Ca enriched surface layer releases Mg ions. *RSC Adv.* **2013**, *3*, 11274–11282. [CrossRef]
42. Song, X.; Tang, W.; Gregurec, D.; Yate, L.; Moya, S.E.; Wang, G. Layered titanates with fibrous nanotopographic features as reservoir for bioactive ions to enhance osteogenesis. *Appl. Surf. Sci.* **2018**, *436*, 653–661. [CrossRef]
43. Inoue, Y.; Uota, M.; Torikai, T.; Watari, T.; Noda, I.; Hotokebuchi, T.; Yada, M. Antibacterial properties of nanostructured silver titanate thin films formed on a titanium plate. *J. Biomed. Mater. Res.* **2010**, *92*, 1171–1180. [CrossRef] [PubMed]
44. Prabu, V.; Karthick, P.; Rajendran, A.; Natarajan, D.; Kiran, M.S.; Pattanayak, D.K. Bioactive Ti alloy with hydrophilicity, antibacterial activity and cytocompatibility. *RSC Adv.* **2015**, *5*, 50767–50777. [CrossRef]
45. Okuzu, Y.; Fujibayashi, S.; Yamaguchi, S.; Yamamoto, K.; Shimizu, T.; Sono, T.; Goto, T.; Ohtsuki, B.; Matsushita, T.; Kokubo, T.; et al. Strontium and magnesium ions released from bioactive titanium metal promote early bone bonding in a rabbit implant model. *Acta Biomater.* **2017**, *63*, 383–392. [CrossRef] [PubMed]
46. Spriano, S.; Yamaguchi, S.; Baino, F.; Ferraris, S. A critical review of multifunctional titanium surfaces: New frontiers for improving osseointegration and host response, avoiding bacteria contamination. *Acta Biomaterialia* **2018**, *79*, 1–22. [CrossRef] [PubMed]
47. Fujibayashi, S.; Neo, M.; Kim, H.-M.; Kokubo, T.; Nakamura, T. A comparative study between in vivo bone ingrowth and in vitro apatite formation on Na_2O-CaO-SiO_2 glasses. *Biomaterials* **2003**, *24*, 1349–1356. [CrossRef]

48. Takemoto, M.; Nakamura, T. *Bioceramics and Their Clinical Applications*; Kokubo, T., Ed.; Woodhead Publishing: Cambridge, UK, 2008; Chapter 8; pp. 183–198.
49. Sila-Asna, M.; Bunyaratvej, A.; Maeda, S.; Kitaguchi, H.; Bunyaratavej, N. Osteoblast differentiation and bone formation gene expression in strontium-inducing bone marrow mesenchymal stem cell. *Kobe J. Med. Sci.* **2007**, *53*, 25–35. [PubMed]

© 2019 by the authors. Licensee MDPI, Basel, Switzerland. This article is an open access article distributed under the terms and conditions of the Creative Commons Attribution (CC BY) license (http://creativecommons.org/licenses/by/4.0/).

Article

The Formation of Microcrystal in Helium Ion Irradiated Aluminum Alloy

Hao Wan [1,*], Shuai Zhao [1], Qi Jin [1], Tingyi Yang [1] and Naichao Si [2]

[1] School of Naval Architecture and Mechanical-Electrical Engineering, Taizhou University, Taizhou 225300, China
[2] School of Materials Science and Engineering, Jiangsu University, Zhenjiang 212013, China
* Correspondence: wanhao@tzu.edu.cn

Received: 8 July 2019; Accepted: 12 August 2019; Published: 15 August 2019

Abstract: A microstructure variation in Al-1060 alloy after helium ion irradiation was revealed by a transmission electron microscope (TEM). The result shows that ion irradiation produced dislocations, dislocation loops, cavities and microcrystals in the irradiated layer. Dislocation-defect interactions were portrayed, especially the pinning effect of a dislocation loop and cavity on moving dislocation. Irradiation-induced stress was recognized as the main factor which impacted on the interaction of defect. Based on the dislocation inhibited with irradiation defects, the mechanism of microcrystal formation was proposed.

Keywords: ion irradiation; dislocation; irradiation defect; microcrystal

1. Introduction

In recent years, microstructure evolution and characteristics of materials subjected to various energetic ion irradiation were investigated [1–3]. When energetic ion impinges on a target surface, the momentum transfer between the incident ion and the target atom will lead to the generation of displacement and rearrangement in an irradiated layer of target. During the irradiation process, interstitial atoms and vacancies are produced continuously and accumulated as cluster defects [4]. Recent studies of irradiation effects in metals have highlighted the important role of primary point defects and clusters in microstructure evolution and property variations of target material.

At a very early stage of irradiation, interstitial atoms have a higher mobility than vacancies, interstitial atoms can accumulate as extrinsic stacking faults, dislocations and interstitial dislocation loops (I-loop). These defects can be observed by a transmission electron microscope (TEM) [5,6]. In Cu^{3+} ion irradiated copper, irradiation-induced defects comprise the stacking fault tetrahedra (SFT) and I-loop, and the defect density can be expressed as a function of the irradiation dose [7].

Similarly, vacancy could aggregate into a vacancy cluster, and a large enough vacancy cluster will collapse into a stacking fault tetrahedra (SFT) or a vacancy type dislocation loop (V-loop) which is intrinsically glissile [8]. On the other hand, a three dimension vacancy cluster was observed in neutron irradiated high purity nickel [9]. During annealing process, these 3D clusters may transform into SFT or V-loop. The formation and evolution of a V-loop in an austenitic stainless steel was in-situ investigated using a laser-equipped high-voltage electron microscope [10].

Another form of vacancy cluster exists as the cavity. Helium-vacancy clusters form at the early stage of irradiation. The accumulation of point defects facilitate the evolution of helium-vacancy clusters into cavities. Using kinetic Monte Carlo simulations, Caturla et al. provided an atomic scale description of this process [11]. Furthermore, glissile dislocation originated from a vacancy cluster provided a channel for atoms transfer which accelerated the rate of void formation [12].

In samples with ion irradiation, it has been known that defects such as dislocations, dislocation loops, SFTs and cavities produced by ion irradiation would alter the performance of an irradiated layer [13–15]. As a commonly used fuel cladding material in a low-temperature reactor, studying the influence of irradiation on microstructure is also beneficial to understand the stress status in the irradiated layer. In our previous work [16,17], irradiation-induced strain variation in the irradiated layer and a morphology change in the surface had been studied. In order to better explain the irradiation-induced tensile stress in an irradiated layer, vacancy cluster was mentioned, while, the configuration of interstitial or vacancy clusters were not studied in detail. The purpose of the present work is to investigate the specific configuration of defect clusters, especially dislocations, dislocation loops and cavities in irradiated samples. However, it has been reported that a lower dose of irradiation would not introduce obvious defects on the target, while an increasing dose of irradiation may raise a higher temperature of the target and result in annealing [18,19]. Synthesizing with our experimental results, the microstructure evolution of irradiated samples was evaluated in this paper. The work provided insight into the defect evolution and interactions, which is helpful to understand the formation mechanism of microcrystals in irradiated samples.

2. Materials and Methods

A commercially pure Al-1060 with a nominal chemical composition ((wt.%): Fe 0.35, Si 0.25, Cu 0.05, Zn 0.05, V 0.05, Mn 0.03, Ti 0.03, Mg 0.03 and Al balance) was studied in this work. Prior to the irradiation experiment, the alloy was cut into square-samples with a cross-section of 10 mm × 10 mm and a thickness of 1 mm by an electrical discharge wire-cutting machine. All samples were previously grounded with an increasing mesh number of SiC sandpapers (from 180 to 1000), finishing with a diamond-alcohol solution polishing compound polish to make a mirror finish. Afterwards, a beam of helium ions (10 $\mu A \cdot cm^{-2}$, 50 keV) were perpendicularly injected into the surface of the target by a MT3-R ion implanter (Beijing BoRuiTianCheng Technology Co., Ltd., Beijing, China) at 300 °C. In [16], a schematic diagram of the home-made MT3-R ion implanter was given. An initial thinning of the samples to sheets with thickness less than 0.1 mm was made by removing materials from the sample undersurface using SiC sandpapers. Then the sheets were punched into discs with diameters of 3 mm. After that, these discs were grounded to a thickness less than 30 µm and thinned in a Gatan-691 precision ion polishing system (Gatan, Inc., Pleasanton, CA, USA). The microstructure of irradiated layer was examined with a JOEL JEM-2100F transmission electron microscope (TEM, JOEL Ltd., Tokyo, Japan) operating at 200 kV.

3. Results

3.1. Microstructure of Un-Irradiated and Irradiated Samples

Figure 1a–d show the bright field TEM images of an un-irradiated sample and samples with 10^{15}, 10^{16} and 10^{17} ions·cm^{-2} of irradiation. A typical microstructure of the Al-1060 alloy shows α-Al matrix with grain boundary (GB) and homogeneously distributed fine precipitates, as shown in Figure 1a. This agrees with the attractive properties of pure aluminum such as good plasticity. The grain size distributions of un-irradiated and irradiated samples were statistical analyzed, as shown in Figure 1e. The average size of a microcrystal in an un-irradiated sample is ~1.37 µm, as shown in Figure 1a. When the alloy was irradiated with 50 keV helium ion, an irradiated layer with a thickness of ~550 nm was produced (calculated by SRIM-2013 software). In the irradiated layer of a 10^{15} ions·cm^{-2} irradiated sample, microcrystals with an average size of ~0.41 µm were found, as shown in Figure 1b. Moreover, smaller grains were found in samples with higher fluence irradiation, as shown in Figure 1c,d. As shown in Figure 1e, the average size of the microcrystals in 10^{16} and 10^{17} ions·cm^{-2} were ~0.26 and ~0.08 µm. The results indicated that the grains were refined in irradiated samples. Since a mass of defects were produced in irradiated samples. These defects would affect the integrity of the crystal structure and induce lattice distortion and stress in irradiated samples. The phenomenon can be revealed by the X-ray diffraction patterns of un-irradiated and irradiated samples that are given in [16]. It can be seen that the (111) peak intensity of the irradiated sample is lower than that of an un-irradiated sample.

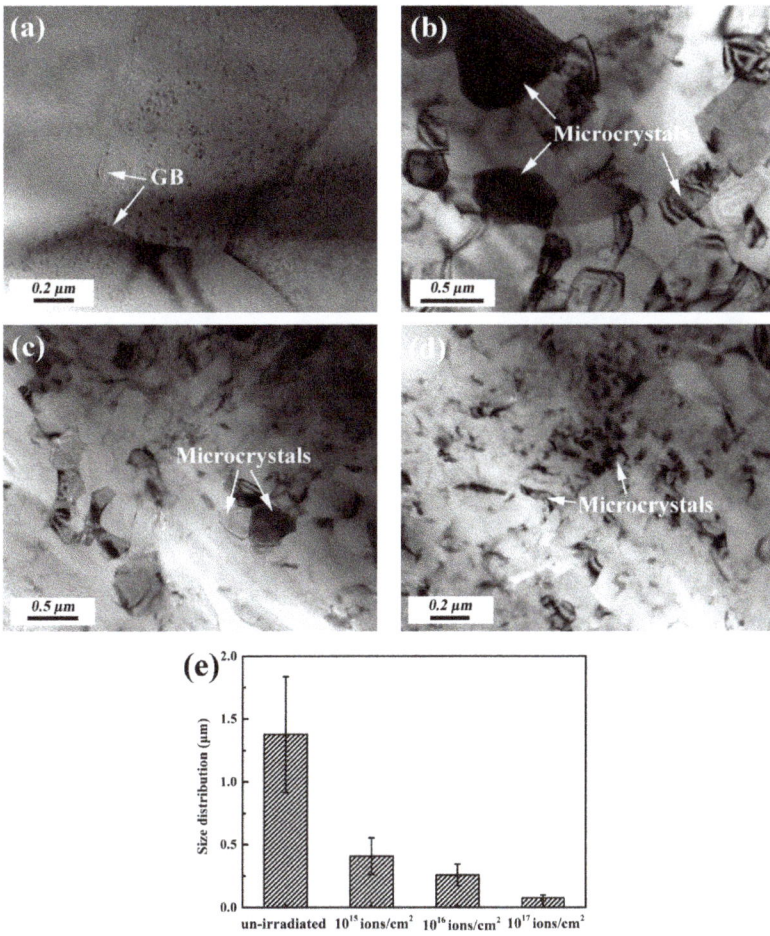

Figure 1. Bright field TEM images of samples without and with irradiation: (**a**) un-irradiated sample; (**b**) 10^{15} ions·cm^{-2}; (**c**) 10^{16} ions·cm^{-2}; and (**d**) 10^{17} ions·cm^{-2}. (**e**) grain size distribution of un-irradiated and irradiated samples.

Figure 2a–e shows the primary microstructure observed in irradiated samples. In comparison with un-irradiated sample (shown in Figure 1a), it can be found that dislocations, dislocation loops and cavities were introduced in the irradiated layer. Figure 2a,b shows the configuration of dislocations and dislocation tangles in the interior of grain. The density of dislocations in this case is much higher than what was shown in the un-irradiated sample. Figure 2c exhibits projected dislocation loops which align along specific directions with different configurations. Patterns of projected dislocation loops change with the variation of angles between habit planes and the view screen, as dislocation loops lay on different habit planes. Rows of dislocation loops exhibiting elliptical contrast were obtained with an average size of ~70 nm. A cavity is another type of irradiation-induced defect, as shown in Figure 2d,e. In overfocus (Figure 2d) and underfocus (Figure 2e) of bright field micrographs, cavities with an average size of ~10 nm can be observed.

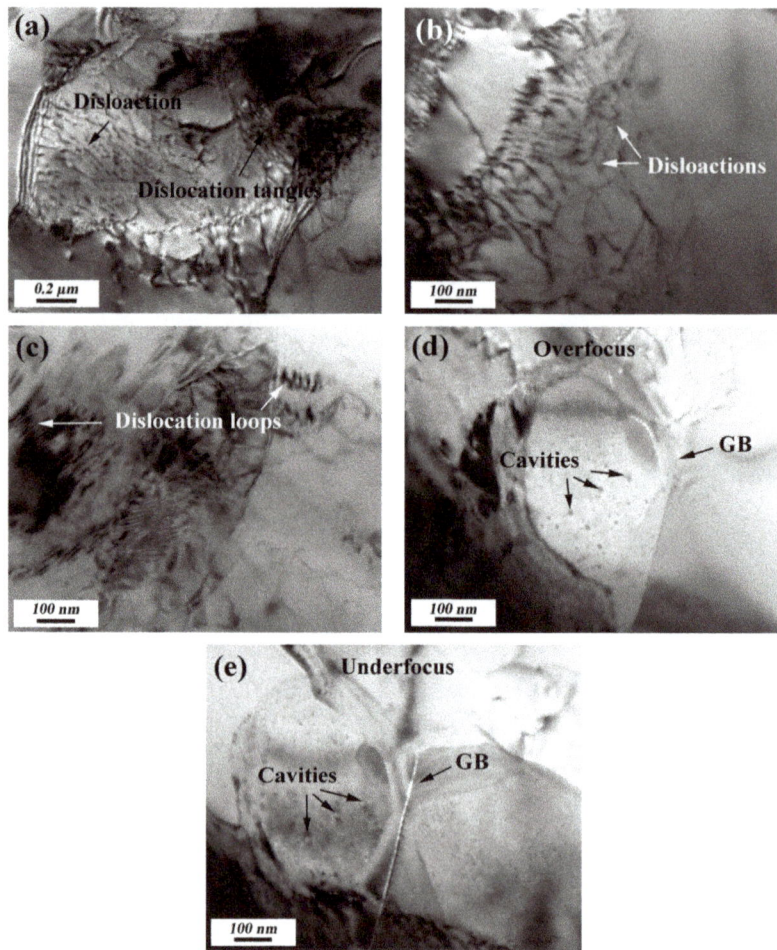

Figure 2. Bright field TEM images of samples with 10^{16} ions·cm^{-2} irradiation: (**a**,**b**) dislocation and dislocation tangle; (**c**) dislocation loop; and (**d**,**e**) cavity.

It is well known that the constant production of self-interstitial atoms (SIAs) and vacancies would introduce lattice distortion, dislocations and other kinds of defects in the irradiated layer of the sample, as shown in Figure 2. In order to reduce the free energy of the matrix, supersaturated vacancies and interstitials tend to accumulate on {111} planes by means of atomic scale diffusion and aggregation. In particular, the accumulated point defects would act as nucleation sites for the formation of dislocation loops and cavities [20,21]. It is noted that transformation of vacancy clusters to V-loops will result in a contraction of the material and induce tensile stress in the direction of the Burgers vector. Similarly, with the formation of I-loops, compression stress is induced. Here, it is suggested that the expansion or shrinkage of the loops are associated with the impact of stress as well as absorption of point defects. Moreover, irradiation-induced stress is recognized as the main factor which facilitates the interaction and growth of defects. The stress is related with lattice distortion and the formation of defects in the irradiated layer.

3.2. Dislocation Configuration in Irradiated Samples

As an increasing number of defects are produced in the surface layer during irradiation, dislocations can be driven by a defects-induced stress field [22]. Figure 3 shows the configuration of dislocation in irradiated samples. Figure 3a indicates the dislocation tangle configuration in the interior of grain. Figure 3b displays the dislocation wall formed in the 10^{16} ions·cm^{-2} irradiated sample. In addition, dislocations are found to pile up and tangle at GB; as shown in Figure 3c,d, the configuration of an array of approaching dislocations pass through GB. In this case, it is considered that the accumulation of dislocations increased the resolved shear stress to GB. In order to relieve the concentration of stress, a special slip system was activated by the GB which facilitated the movement of dislocations. In fcc metals, {111} semi-coherent interfaces contain either three sets of Shockley partial dislocations or three sets of full dislocations, depending on the stacking fault energy [23]. The high stacking fault energy of aluminum promotes the nucleation of partial dislocation during movement, which removes the stacking fault defects on glide planes [24]. That is why no obvious stacking fault defects can be observed in the irradiated sample.

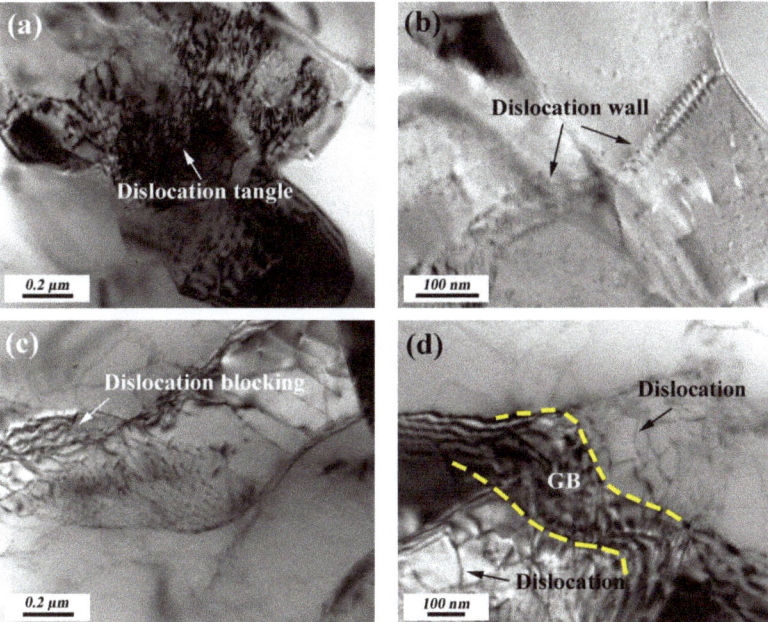

Figure 3. Bright field TEM images of dislocation configuration in 10^{16} ions·cm^{-2} irradiated sample: (**a**) dislocation tangle; (**b**) dislocation wall; (**c**) dislocation blocking; and (**d**) dislocation pass through GB.

3.3. Dislocation-Defect Interaction in Irradiated Samples

What is noteworthy is that, irradiation-induced defects were acting as obstacles for moving dislocation. It has been known that dislocation can be easily driven at higher temperatures in a sample existing with applied stress. In this case, the interaction of moving dislocations with obstacles deserved more attention. As shown in Figure 3, the dislocation–dislocation interaction may form a dislocation tangle, dislocation wall, etc. Figure 4 shows the configuration of a moving dislocation meeting with obstacles such as a dislocation loop and a cavity in irradiated samples. Figure 4a shows that dislocations bow out as a result of inhibition of dislocation loops. Figure 4b shows the configuration of moving dislocations pinned with dislocation loops. As the number of dislocations increase at pinning sites, a change of strain field can be inferred from the distance variation of dislocations. In practice, loops with a larger size can block the motion of dislocations more effectively.

Furthermore, the rows of dislocation loops along specific directions revealed in Figures 1c and 4c deserve more attention. As pre-existing crystal defects such as dislocations and GBs would act as sinks for trapping or recombining irradiation-induced interstitial atoms [25]. Moreover, interstitial atoms always have lower migration ability and formation energy than vacancies. Hence, a greater probability of capturing interstitial atoms would occur at sinks. As interstitial atoms captured with trapping sites, a favorable condition is created for the formation of vacancy clusters [26]. As shown in Figure 4b,c, dislocation loops are favored to form in the vicinity of GBs [27]. In addition, a group of dislocation loops have a better pinning effect than single ones, which facilitate the formation of microcrystals in irradiated samples.

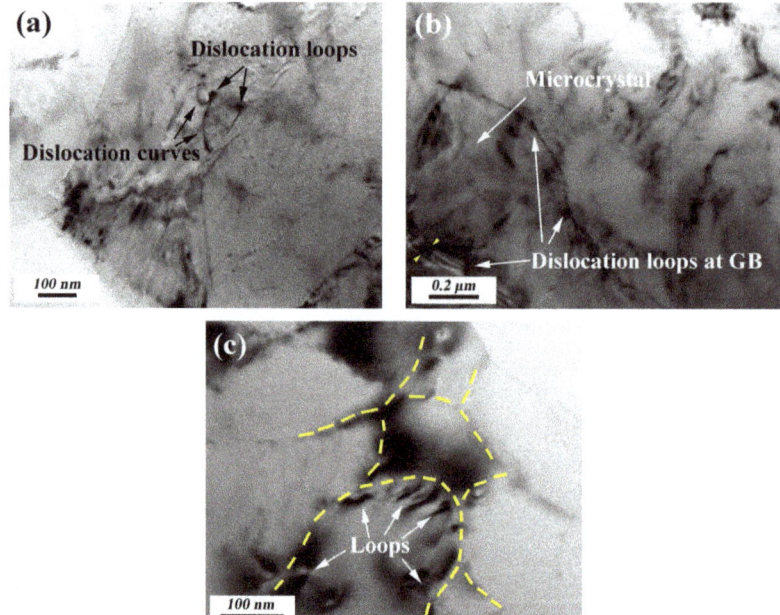

Figure 4. Interaction of dislocation with dislocation loop: (**a**) dislocations pinned with dislocation loops, 10^{15} ions·cm^{-2}; (**b**,**c**) dislocation loops near GB, 10^{16} ions·cm^{-2}.

Similarly, cavities can also perform as effective barriers in the process of dislocation movement, as shown in Figure 4b. Vacancy clusters can be present in the form of cavities in irradiated samples, as observed in Figure 1d,e. During helium ion irradiation, vacancies or vacancy clusters can trap helium atoms to form He$_n$V$_m$ type vacancy clusters [28]. Attributed to the helium induced pressurization inside, these He$_n$V$_m$ type vacancy clusters exhibited more stability than empty vacancy clusters [29]. Moreover, irradiation enhanced the diffusion ability of helium atoms and vacancy, which accelerated He$_n$V$_m$ complex nucleation and coalescence. Then, cavities formed and grew into bigger ones. Cavities formed in irradiated samples have the effect of blocking and pinning dislocation, as shown in Figure 5b. Figure 5b shows the tendency of small cavity coalescence into bigger ones. Similarly, with V-loops, cavities have drawing force on surrounding material, which would also introduce tensile stress in the irradiated layer. In addition, bigger cavities can enhance the barrier effect on moving dislocations, which will take longer time to climb over the bigger ones. Analogous conclusions were revealed in fcc metals simulated by a concurrent atomistic-continuum method, dislocations were also found to bow as a result of pinning on the original glide planes due to cavity strengthening [30,31]. Moreover, when cavities appear in a line, it can be ascertained that an ideal site was developed for the production of microcrystal boundary, as shown in Figure 5a. Accordingly, dislocation-defect interactions facilitated the forming of microcrystal in irradiated samples. In conclusion, irradiation induced basic defects (vacancies and interstitials) in irradiated samples. The diffusion and aggregation of point defects resulted in the forming of structural defects. Moreover, the defect–defect interactions facilitated the evolution of microstructure (involving microcrystal forming) in irradiated samples.

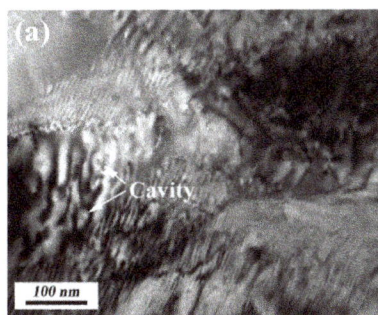 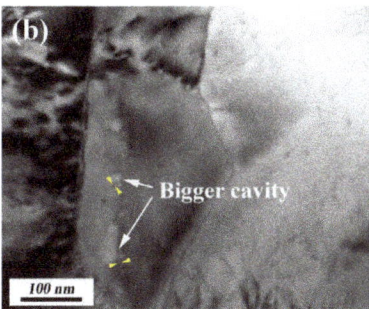

Figure 5. (**a**) dislocation-cavity interaction, 10^{16} ions·cm^{-2}; (**b**) bigger cavity, 10^{16} ions·cm^{-2}.

4. Conclusions

Microstructure variations of helium ion irradiated Al-1060 alloys were explored in this work. Irradiation-induced supersaturated point defects led to the formation of irradiation defects such as dislocations, dislocation loops and cavities in irradiated samples. Attributing to the dislocation-defects interactions, microcrystals were formed in the irradiated layer. Dislocation loops may expand or shrink by means of absorbing point defects. In addition, dislocation loops prefer to form in the vicinity of GBs due to the sink effect and high stacking fault of aluminum. Helium can stabilize the nucleation and growth of vacancy clusters. Meanwhile, cavities can grow by absorbing a vacancy cluster or a small cavity from an adjacent region. Both dislocation loops and cavities with bigger sizes would have a better barrier effect than smaller ones. In addition, the interaction between dislocation and irradiation defects was the primary mechanism for the production of microcrystal in irradiated samples.

Author Contributions: Conceptualization, H.W.; Methodology, H.W. and N.S.; Formal Analysis, H.W.; Investigation, H.W; Writing—Original Draft Preparation, H.W.; Writing—Review and Editing, H.W., S.Z., Q.J. and T.Y.

Funding: This work was supported by the Scientific Research Foundation of Taizhou University for the Introduction of Talents (TZXY2017QDJJ014), the sixteenth "Six talent peaks" of high-level talent selection and a training project of Jiangsu Province (XCL-265), the Natural Science Foundation of the Jiangsu Higher Education Institutions of China (18KJD430007), the college students' innovation and entrepreneurship training project of Jiangsu province (201912917035Y and 201912917029Y), and the Fifth "311 High-level Talents Training Project" of Taizhou City.

Conflicts of Interest: The authors declare no conflict of interest.

References

1. Yi, X.; Jenkins, M.L.; Hattar, K.; Philip, D.E.; Steve, G.R. Characterisation of radiation damage in W and W-based alloys from 2 MeV self-ion near-bulk implantations. *Acta Mater.* **2015**, *92*, 163–177. [CrossRef]
2. Wang, C.; Shan, D.; Guo, B.; Xue, J.; Zhang, H. Effect of nano-crystals at surfaces induced by ion beam irradiation on the tribological behaviour in microforming. *Vacuum* **2013**, *89*, 267–270. [CrossRef]
3. Gao, J.; Yabuuchi, K.; Kimura, A. Ion-irradiation hardening and microstructural evolution in F82H and ferritic alloys. *J. Nucl. Mater.* **2019**, *515*, 294–302. [CrossRef]

4. Wan, F.; Zhan, Q.; Long, Y.; Yang, S.; Zhang, G.; Du, Y.; Jiao, Z.; Ohnuki, S. The behavior of vacancy-type dislocation loops under electron irradiation in iron. *J. Nucl. Mater.* **2014**, *455*, 253–257. [CrossRef]
5. Khan, A.K.; Yao, Z.; Daymond, M.R.; Holt, R.A. Effect of foil orientation on damage accumulation during irradiation in magnesium and annealing response of dislocation loops. *J. Nucl. Mater.* **2012**, *423*, 132–141. [CrossRef]
6. Zhang, H.K.; Yao, Z.W.; Morin, G.; Griffiths, M. TEM characterization of in-reactor neutron irradiated CANDU spacer material Inconel X-750. *J. Nucl. Mater.* **2014**, *451*, 88–96. [CrossRef]
7. Li, N.; Hattar, K.; Misra, A. In situ probing of the evolution of irradiation-induced defects in copper. *J. Nucl. Mater.* **2013**, *439*, 185–191. [CrossRef]
8. Satoh, Y.; Matsuda, Y.; Yoshiie, T.; Kawai, M.; Matsumura, H.; Iwase, H.; Abe, H.; Kim, S.W.; Matsunaga, T. Defect clusters formed from large collision cascades in fcc metals irradiated with spallation neutrons. *J. Nucl. Mater.* **2013**, *442*, S768–S772. [CrossRef]
9. Druzhkov, A.P.; Arbuzov, V.L.; Perminv, D.A. The effect of neutron irradiation dose on vacancy defect accumulation and annealing in pure nickel. *J. Nucl. Mater.* **2012**, *421*, 58–63. [CrossRef]
10. Yang, Z.B.; Watanabe, S. Dislocation loop formation under various irradiations of laser and/or electron beams. *Acta Mater.* **2013**, *61*, 2966–2972. [CrossRef]
11. Caturla, M.J.; Rybia, T.D.; Fluss, M. Modeling microstructure evolution of fcc metals under irradiation in the presence of He. *J. Nucl. Mater.* **2003**, *323*, 163–168. [CrossRef]
12. Di, S.; Yao, Z.W.; Daymond, M.R.; Zu, X.T.; Peng, S.M.; Gao, F. Dislocation-accelerated void formation under irradiation in zirconium. *Acta Mater.* **2015**, *82*, 94–99. [CrossRef]
13. Draganski, M.A.; Finkman, E.; Gibson, B.C.; Fairchild, B.A.; Ganesan, K.; Nabatova-Gabain, N.; Tomljenovic-Hanic, S.; Greentree, A.D.; Prawer, S. The effect of gallium implantation on the optical properties of diamond. *Diam. Relat. Mater.* **2013**, *35*, 47–52. [CrossRef]
14. Wan, H.; Ding, Z.; Wang, J.; Yin, Y.; Guo, Q.; Gong, Y.; Zhao, Z.; Yao, X. Effects of helium ion irradiation on the high temperature oxidation resistance of Inconel 718 alloy. *Surf. Coat. Technol.* **2019**, *363*, 34–42. [CrossRef]
15. Jayalakshmi, G.; Saravanan, K.; Balakumar, S.; Balasubramanian, T. Swift heavy ion induced modifications in structural, optical & magnetic properties of pure and V doped ZnO films. *Vacuum* **2013**, *95*, 66–70. [CrossRef]
16. Wan, H.; Si, N.; Chen, K.; Wang, Q. Strain and structure order variation of pure aluminum due to helium irradiation. *RSC Adv.* **2015**, *5*, 75390–75394. [CrossRef]
17. Wan, H.; Si, N.; Wang, Q.; Zhao, Z. Morphology variation, composition alteration and microstructure changes in ion-irradiated 1060 aluminum alloy. *Mater. Res. Express* **2018**, *5*, 026501. [CrossRef]
18. Martin, G.; Garcia, P.; Sabathier, C.; Van Brutzel, L.; Dorado, B.; Garrido, F.; Maillard, S. Irradiation-induced heterogeneous nucleation in uranium dioxide. *Phys. Lett. A* **2010**, *374*, 3038–3041. [CrossRef]
19. Jin, S.; Guo, L.; Yang, Z.; Fu, D.; Liu, C.; Xiao, W.; Tang, R.; Liu, F.; Qiao, Y. Microstructural evolution in nickel alloy C-276 after Ar^+ ion irradiation. *Nucl. Instrum. Method Phys. Res. Sect. B* **2011**, *269*, 209–215. [CrossRef]
20. Zinkle, S.J. Effect of H and He irradiation on cavity formation and blistering in ceramics. *Nucl. Instrum. Method Phys. Res. Sect. B* **2012**, *286*, 4–19. [CrossRef]
21. Liu, L.; Liu, Q.; Wang, Z.; Tang, Z. Interstitial clusters on $\Sigma = 11(113)$ grain boundary in copper: Geometric structure, stability, and ability to annihilate vacancies. *Phys. Lett. A* **2016**, *380*, 621–627. [CrossRef]
22. Ke, J.H.; Boyne, A.; Wang, Y.; Kao, C.R. Phase field microelasticity model of dislocation climb: Methodology and applications. *Acta Mater.* **2014**, *79*, 396–410. [CrossRef]
23. Shao, S.; Wang, J.; Misra, A.; Hoagland, R.G. Spiral patterns of dislocations at nodes in (111) semi-coherent FCC interfaces. Scientific reports. *Sci. Rep.* **2013**, *3*, 2448. [CrossRef] [PubMed]
24. Terentyev, D.; Bakaev, A.; Osetsky, Y.N. Interaction of dislocations with Frank loops in Fe–Ni alloys and pure Ni: An MD study. *J. Nucl. Mater.* **2013**, *442*, S628–S632. [CrossRef]
25. Wan, H.; Si, N.; Zhao, Z.; Wang, J.; Zhang, Y. Study of irradiation induced surface pattern and structural changes in Inconel 718 alloy. *Mater. Res. Express* **2018**, *5*, 056503. [CrossRef]
26. Zhang, H.; Yao, Z.; Daymond, M.R. Cavity morphology in a Ni based superalloy under heavy ion irradiation with hot pre-injected helium. II. *J. Appl. Phys.* **2014**, *115*, 103509. [CrossRef]
27. Mizuno, K.; Morikawa, K.; Okamoto, H.; Hashimoto, E. Row of dislocation loops as a vacancy source in ultrahigh-purity aluminum single crystals with a low dislocation density. *Trans. Mater. Res. Soc. Jpn.* **2014**, *3*, 169–172. [CrossRef]

28. Zhang, H.; Yao, Z.; Daymond, M.R. Cavity morphology in a Ni based superalloy under heavy ion irradiation with cold pre-injected helium. I. *J. Appl. Phys.* **2014**, *115*, 103508. [CrossRef]
29. Ou, X.; Anwand, W.; Kögler, R.; Zhou, H.-B.; Richter, A. The role of helium implantation induced vacancy defect on hardening of tungsten. *J. Appl. Phys.* **2014**, *115*, 123521. [CrossRef]
30. Wan, H.; Si, N.; Wang, Q.; Chen, K.; Liu, G. Surface morphology alteration, microstructure variation and dislocation-precipitate interactions of Inconel 718 due to helium ions irradiation. *Mater. Charact.* **2017**, *127*, 95–103. [CrossRef]
31. Xiong, L.; Xu, S.; McDowell, D.L.; Chen, Y. Concurrent atomistic–continuum simulations of dislocation–void interactions in fcc crystals. *Int. J. Plast.* **2015**, *65*, 33–42. [CrossRef]

 © 2019 by the authors. Licensee MDPI, Basel, Switzerland. This article is an open access article distributed under the terms and conditions of the Creative Commons Attribution (CC BY) license (http://creativecommons.org/licenses/by/4.0/).

Article

The Effect of Heat Treatment on Properties of Ni–P Coatings Deposited on a AZ91 Magnesium Alloy

Martin Buchtík *, Michaela Krystýnová, Jiří Másilko and Jaromír Wasserbauer

Materials Research Centre, Faculty of Chemistry, Brno University of Technology, Purkyňova 464/118, 61200 Brno, Czech Republic
* Correspondence: xcbuchtik@fch.vut.cz; Tel.: +420-736-445-019

Received: 10 June 2019; Accepted: 19 July 2019; Published: 23 July 2019

Abstract: The present study reports the effect of phosphorus content in deposited electroless nickel (Ni–P) coatings, the heat treatment on the microhardness and its microstructural characteristics, and the influence of the temperature on the microstructure of the Mg alloy substrate during the heat treatment. The deposition of Ni–P coatings was carried out in the electroless nickel bath, and the resulting P content ranged from 5.2 to 10.8 wt.%. Prepared samples were heat-treated in the muffle furnace at 400 °C for 1 h after the coating deposition. The cooling of the samples to room temperature was proceeded in the air. For as-deposited and heat-treated samples, it was determined that with the increasing P content, the microhardness was decreasing. This may be caused by the changes in the structure of the Ni–P coating. The X-ray diffraction patterns of the as-deposited Ni–P coatings showed that the microstructure changed their nature from crystalline to amorphous with the increasing P content. The heat treatment of prepared samples led to the significant increase of microhardness of Ni–P coatings. All the heat-treated samples showed the crystalline character, regardless of the P content and the presence of hard Ni_3P phase, which can have a positive effect on the increase of microhardness. The metallographic analysis showed changes of substrate microstructure after the heat treatment. The prepared coatings were uniform and with no visible defects.

Keywords: Ni–P coatings; Ni_3P phase; Mg alloys; AZ91; heat treatment; microhardness; crystallite size

1. Introduction

Magnesium alloys are the lightest structural metallic materials [1,2]. Due to their exceptional properties, such as a low density, stiffness, specific strength, good castability, and machinability, they are desirable in various industries [3–5]. One of the biggest limits for the widespread use of magnesium alloys is their poor corrosion, wear resistance, and low hardness [1,5–7]. These problems are often resolved by means of surface coatings. Electroless nickel (Ni–P) deposition seems to be an appropriate variant to protect magnesium alloy substrates [8]. Electroless Ni–P coatings are mainly used due to their excellent corrosion resistance, high hardness, and wear resistance. However, properties of Ni–P coatings are strongly dependent on their chemical composition, i.e., the phosphorus (P) content in the coating [9]. In terms of the chemical composition, electroless Ni–P coatings can be divided into three groups: Low phosphorus (1–5 wt.% of P), medium phosphorus (6–9 wt.% of P), and high phosphorus (10–13 wt.% of P) [9–11]. Low phosphorus Ni–P coatings are predominantly crystalline and less corrosion resistant compared to the medium and high P coatings. They are characteristic with a high hardness and good mechanical and tribological properties. The crystalline character of the low phosphorus coatings indicates that the number of phosphorus atoms in interstitial positions is not sufficient for the distortion of the nickel lattice [11,12].

High phosphorus Ni–P coatings are known for excellent corrosion resistance due to their amorphous microstructure [11–13].

Duncan [11] stated that Ni–P coating is in non-equilibrium state after deposition. Ni–P coating is formed by a crystalline solid solution of P in Ni, called the β phase (low phosphorus), the total amorphous γ phase, which exists between 11–15 wt.% P (high phosphorus), or the mixture of β + γ phase (medium phosphorus). These metastable phases are characterized by decomposition reactions during the heat treatment to form the equilibrium α phase (solid solution of P in Ni) and Ni$_3$P phase. Crystalline nickel (the α phase) and the Ni$_3$P phase are only stable products after the heat treatment. These stable phases begin to form from 300 °C. The optimal temperature range for the heat treatment of the Ni–P coatings is within the temperature range of 300 to 400 °C. Riedel [10] stated that it is advisable to perform heat treatment at 400 °C for 1 h to achieve the maximum hardness of Ni–P coatings. The increase in precipitate size and coating grain coarsening was observed at applied temperatures higher than 400 °C and longer treating times during the heat treatment process, regardless of the P content [14,15].

A suitable heat treatment process can result in an increase in the coating hardness, up to 1300 HV. This is because of the recrystallization of a non-equilibrium β phase (low phosphorus), an amorphous γ phase (high phosphorus), or their mixture (medium phosphorus) into the equilibrium crystalline α phase, with a simultaneous precipitation of the hard intermediate Ni$_3$P phase [10,14].

Kumar [16] reported that the crystallite size of Ni changes with the increasing heat treatment temperature. From the room temperature to 100 °C, there was only a negligible change in the Ni crystallite size. Between 100 and 300 °C, the increase in crystallite size of Ni was evident due to the arrangement of Ni atoms in the lattice. However, no formation of any intermediate precipitate particles was detected. Authors also listed that the disappearance of the amorphous phase was observed at 330 °C, what indicates the complete crystallization of the microstructure. At a temperature above 300 °C, a significant increase in crystallite size was observed, probably due to the formation of Ni$_3$P phase particles.

Most of the published studies are focused on the influence of heat treatment and P content in Ni–P coatings deposited on the steels. However, the heat treatment of Ni–P coatings to achieve the maximum hardness is performed in the temperature of 400 °C for 1 h. This temperature does not influence the microstructure of steels but may have significant effect on magnesium alloys. Therefore, this study deals with the effects of the heat treatment of Ni–P coatings with the various P content deposited on a AZ91 magnesium alloy. The microstructure of a AZ91 alloy and the characterization of Ni–P coatings, such as microhardness, phase composition, and crystallite size, were evaluated before and after the heat treatment.

2. Materials and Methods

Samples of a cast AZ91 magnesium alloy with dimensions of 30 × 30 × 7 mm^3 were chosen as substrates for the electroless deposition of Ni–P coatings. The elemental composition of the AZ91 alloy, analyzed using the glow-discharge optical emission spectroscopy (GDOES) Spectrumat GDS 750 (Spectruma Analytik GmbH, Hof, Germany), is listed in Table 1. To obtain an appropriate surface, the samples of the Mg alloy were ground using no. 1200 SiC paper before the pre-treatment process. During the pre-treatment process, ground samples were degreased in an alkali bath and then pickled in an acid-pickling bath to activate the surface. After each step of the pre-treatment, samples were rinsed in distilled water and isopropyl alcohol and then dried in hot air. The deposition of Ni–P coatings was carried out in the electroless nickel bath with different Ni^{2+}/H$_2$PO$_2^-$ ratios. Individual ratios of Ni^{2+}/H$_2$PO$_2^-$ were set at 0.1, 0.2, 0.3, 0.45, and 0.75. The chemical composition was characterized using a Zeiss EVO LS-10 (Carl Zeiss Ltd., Cambridge, UK) scanning electron microscope (SEM) with energy-dispersive spectroscopy (EDS) Oxford Instruments Xmax 80 mm^2 detector (Oxford Instruments plc, Abingdon, UK) and the AZtec software (version 2.4).

Table 1. Elemental composition of the AZ91 Mg alloy, glow-discharge optical emission spectroscopy (GDOES).

Element	Al	Zn	Cu	Mn	Si	Fe	Ni	Zr	Mg
Content [wt.%]	8.80	0.81	0.00	0.32	0.01	0.004	0.00	0.01	Bal.

Prepared samples were heat-treated in the muffle furnace LAC LM07 (LAC, s.r.o., Židlochovice, Czech Republic) at 400 °C for 1 h after the coating deposition. The cooling of the samples to room temperature was proceeded in the air.

The microstructure of the Ni–P coatings and AZ91 magnesium alloy was characterized using an Axio Observer Z1m (ZEISS) light microscope and a Zeiss EVO LS-10 scanning electron microscope.

The microhardness of the deposited Ni–P coatings was measured using a LECO AMH55 microhardness tester (Saint Joseph, MO, USA). The microhardness was performed and evaluated according to the ASTM E384 standard. The microhardness was measured from the perpendicular cut. The samples were ground and polished using a Tegramin-25 (Struers) automatic grinder with a special holder for preparation of planar specimens. The final step was polishing, using diamond paste with 0.25 μm particle size. Iso-propanol was used as a lubricant. The Vickers method was used with the applied load of 25 g for 10 s. The microhardness value was determined from 10 values.

For the determination and characterization of the Ni–P coatings phase composition, the coatings were mechanically separated from the substrate, milled, and analyzed in the powder form using the Scherrer method. The analysis was performed on an Empyrean (Panalytical) X-Ray diffraction spectrometer with Cu-anode ($\lambda K\alpha_1$ = 0.15406 nm, $\lambda K\alpha_2$ = 0.15444 nm) at room temperature. The scan step size was set up at 0.013°. The obtained data were processed using High Score Plus software. The crystallite size of Ni and Ni_3P was calculated from the full width half maximum (FWHM) according to the Scherrer equation [17] (Equation (1)):

$$\tau = \frac{K \cdot \lambda}{\beta_{1/2} \cdot \cos\theta} \quad (1)$$

where τ is the crystallite size, λ is the X-ray wavelength, $\beta_{1/2}$ is the peak extension at half of the maximum intensity (FWHM), θ is the diffraction Bragg's angle, and K is the particles shape factor (Scherrer constant) depending on the shape of the crystallites. K, ranging from 0.62 to 2.08, is usually close to 1. For perfectly rounded crystals, K is equal to 0.89.

3. Results and Discussion

3.1. Microstructure and Chemical Composition

Table 2 shows the results of the chemical composition of the as-deposited and heat-treated Ni–P coatings, deposited on the AZ91 alloy with different $Ni^{2+}/H_2PO_2^-$ ratios in the electroless nickel bath. The average phosphorus content in as-deposited and heat-treated coatings was similar for the same $Ni^{2+}/H_2PO_2^-$ ratios, therefore there is only one value for each $Ni^{2+}/H_2PO_2^-$ ratio. The average P content ranges from approximately 5 wt.% to 11 wt.% of P, both for as-deposited and heat-treated coatings.

Table 2. The phosphorus content of electroless nickel (Ni–P) as-deposited and heat-treated coatings in dependence on the $Ni^{2+}/H_2PO_2^-$ ratios, energy-dispersive spectroscopy (EDS).

$Ni^{2+}/H_2PO_2^-$ Ratio	P Content [wt.%]
0.75	5.2 ± 0.2
0.45	5.5 ± 0.1
0.3	7.4 ± 0.1
0.2	10.1 ± 0.2
0.1	10.8 ± 0.1

Figure 1 shows the microstructure of as-deposited and heat-treated Ni–P. The microstructures of as-deposited and heat-treated Ni–P coatings were similar, regardless of the chemical composition. The average thickness of all coatings was approximately 30 µm. The coating was uniform without structural defects and there was no undesirable interlayer between the magnesium alloy substrate and Ni–P coating. The heat treatment did not affect the thickness or overall chemical composition of the deposited Ni–P coatings.

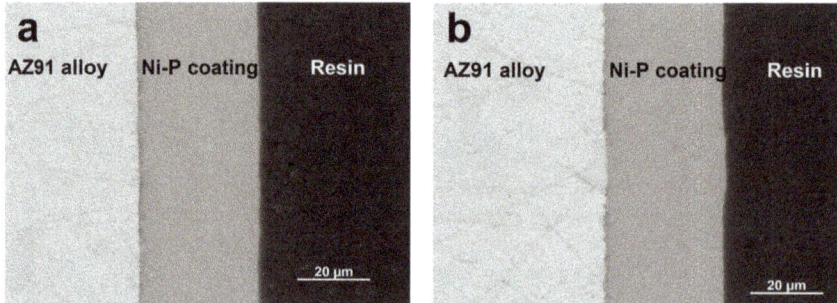

Figure 1. Microstructure of Ni–P coatings with 7.4 wt.% P (**a**) as-deposited; (**b**) heat-treated.

As shown in Figure 2a, the microstructure of the cast AZ91 magnesium alloy consists of (1) α solid solution of Al in Mg, (2) discontinuous precipitates of intermetallic $Mg_{17}Al_{12}$-$β_D$ phase, and (3) eutectic α + β [1,2].

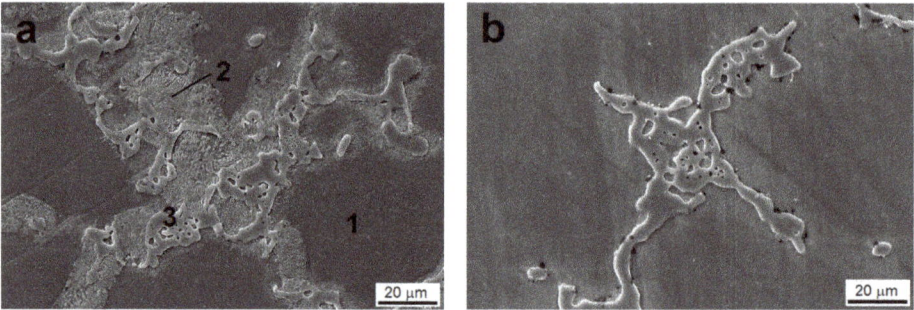

Figure 2. Microstructure of the AZ91 Mg alloy (**a**) as-cast, (**b**) heat-treated at 400 °C for 1 h.

In the case of heat-treated samples, the microstructure of the AZ91 alloy was changed. As seen in Figure 2b, the presence of the discontinuous precipitate of the $Mg_{17}Al_{12}$ phase was not observed. This finding can be explained by the fact that the discontinuous precipitate was dissolved in α solid solution of Al in Mg during the heat treatment at 400 °C for 1 h. Due to the fast cooling in the air after being removed from the furnace, discontinuous precipitates of the $Mg_{17}Al_{12}$-$β_D$ phase were not present [2]. However, the $Mg_{17}Al_{12}$-β phase and eutectic α + β was still observed in the microstructure.

Because of the dissolution of the discontinuous precipitates of the $Mg_{17}Al_{12}$-$β_D$ phase at 400 °C, the content of Al in the α solid solution increased, which may lead to improvement of some mechanical properties due to the solid solution strengthening.

3.2. Microhardness of Ni–P Coatings

As stated in the literature [10,18], low-phosphorus Ni–P coatings are crystalline, medium-phosphorus coatings are microcrystalline, and high-phosphorus Ni–P coatings are amorphous. The microstructure of deposited Ni–P coatings strictly affects their properties [12,13]. In general, the microhardness decreases with the increasing P content.

From the results of microhardness, the measurement can be stated that the highest microhardness value was observed in the case of the Ni–P coating with the lowest P. With the increasing P content, the microhardness decreased in the case of both the as-deposited and in the heat-treated coatings (Figure 3), which is in correlation with the literature [12,13,19]. Ashtiani et al. [19] reported that in the case of the Ni–P coatings heat-treated at 400 °C, the microhardness decreased with the increasing P content, which is in agreement with presented data. Deposited Ni–P coating with 9.35 wt.% of P reached the microhardness of 970 HV 50 gf. Meanwhile, the coatings with 10.31 and 11.45 wt.% of P reached the microhardness of 856 HV 50 gf and 788 HV 50 gf, respectively.

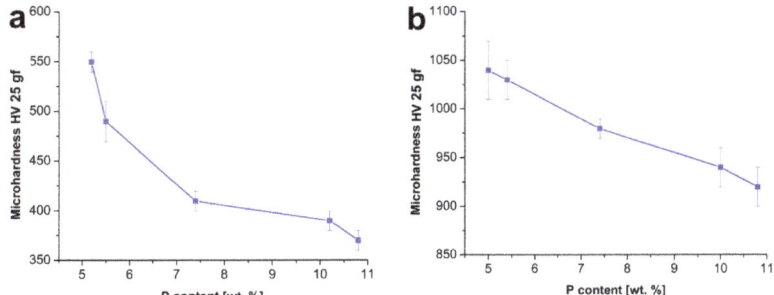

Figure 3. Microhardness dependence on the phosphorus content of Ni–P coatings, (**a**) as-deposited, (**b**) heat-treated.

Figure 3b shows that the heat-treated Ni–P coatings had higher value of microhardness when compared to the as-deposited coatings with the same P content. During the heat treatment, all the coatings (depending on the P content) became more crystalline due to the rearrangement of the structure and the transformation of the non-equilibrium solid solution P in the Ni-β phase (low phosphorus), total amorphous γ phase (high phosphorus), or their mixture (medium phosphorus) to equilibrium crystalline solid solution P in Ni-α phase. Simultaneously, heat treating lead to the formation of the hard body centered tetragonal Ni_3P phase. The presence of Ni_3P results in an overall increase of the microhardness of the coatings [9,10]. The presence of the Ni_3P phase in the heat-treated coatings was confirmed by XRD analysis.

3.3. Phase Analysis of Ni–P Coatings

The XRD patterns corresponding to the individual measurements representing Ni–P coatings are provided in Figure 4. As seen in the Figure 4a, the peak corresponding to the fcc nickel crystal lattice (1 1 1) can be observed near the diffraction angle $2\theta \approx 45°$. A broad peak corresponding to the Ni diffraction was observed in the case of high-phosphorus coatings, and with the decreasing P content, the peak of Ni became sharper. This effect indicates a more ordered internal microstructure [18]. The highest intensity of Ni diffraction was measured for the Ni–P coating with 5.5 ± 0.1 wt.% of P. Meanwhile, the lowest intensity and the broadest peak was observed in the case of the Ni–P coating with 10.8 ± 0.1 wt.% of P.

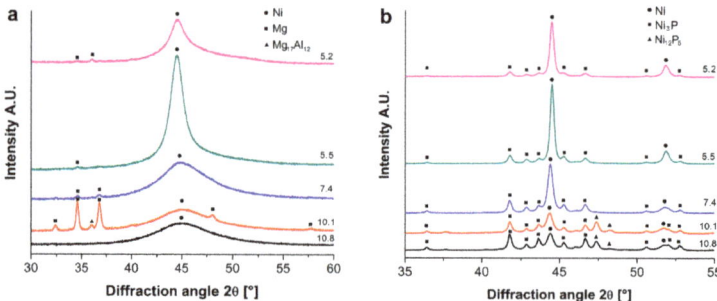

Figure 4. XRD patterns of (**a**) as-deposited and (**b**) heat-treated Ni–P coatings with different phosphorus content.

Except for the diffraction of Ni, there were clear diffractions between 30°–40° and around the angle 2θ ≈ 48°. Gu [20] listed that these XRD peaks correspond to the primary α-Mg phase and the $Mg_{17}Al_{12}$ phase. The α-Mg phase was also detected in the work of Hu [21].

The presence of the phase particles can be explained by the fact that there was a joint separation of the Mg alloy together with the Ni–P coating during the mechanical separation. However, the presence of these phases in the tested powder did not affect the microstructural changes observed in deposited Ni–P coatings.

Figure 4b shows the patterns of heat-treated Ni–P coatings with different P content. The diffraction of Ni (1 1 1) can be seen near the diffraction angle 2θ ≈ 44.4°. Another diffraction of Ni (2 0 0) can be seen at 2θ ≈ 51.8°. For both of the Ni diffractions, it can be observed that their intensity increases and peaks become more sharp with the decreasing P content (except the P content 5.2 ± 0.2 wt.%, which is slightly lower than the peak for the Ni–P coating with the P content 5.5 ± 0.1 wt.%).

Figure 4b shows that the presence of the Ni_3P stable phase was obvious for all the coatings and the intensity of the Ni_3P phase increased with the increasing P content. From patterns shown in Figure 4b, it is evident that in the case of low-phosphorus Ni–P coatings, the Ni phase crystallizes more (the peak of Ni is sharper and with higher intensity) when compared to the high-phosphorus Ni–P coatings. On the other hand, the Ni_3P phase precipitated and grew more in the case of the high-phosphorus Ni–P coating. The presence of the $Ni_{12}P_5$ metastable phase was observed in the case of high-phosphorus Ni–P coatings (10.2 and 10.8 wt.% of P) around the diffraction angle 2θ ≈ 47° to 48°. According to the literature [22,23], the $Ni_{12}P_5$ metastable phase should completely disappear around the temperature of 350 °C. However, Keong [24] showed that the $Ni_{12}P_5$ phase may still be present at 400 °C. The presence of this phase could be caused by the incomplete transformation from the originally amorphous matrix to the mixture of crystalline Ni and the Ni_3P stable phase.

3.4. Crystallite Size

Figure 5 shows the effect of the phosphorus content on the Ni crystallite size in as-deposited coatings and heat-treated coatings. Only one diffraction plane Ni (1 1 1) was observed (see Figure 4a) and two diffraction planes (1 1 1) and (2 0 0) of Ni were observed (Figure 4b) in the case of as-deposited and heat-treated coatings by XRD, respectively. Figure 5 shows that with the increasing P content, the crystallite size of nickel decreases, both in the case of the as-deposited and heat-treated coatings. This fact can be explained due to the increasing lattice disorder (a greater proportion of the amorphous phase) with the increasing P content in the Ni–P matrix [12,24].

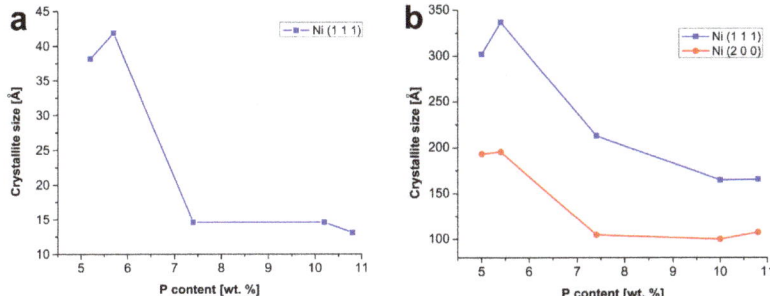

Figure 5. The effect of Ni crystallite size on the P content in (**a**) as-deposited Ni–P coatings and (**b**) heat-treated Ni–P coatings.

As can be seen in Figure 5a, the crystallites of Ni in low-phosphorus as-deposited Ni–P coatings reached approximately 40 Å, whereas the crystallites of Ni in the Ni–P coating with 10.8 wt.% of P reached the size of 13.1 Å. After the heat treatment, the crystallite size of Ni substantially increased to more than 300 Å in the case of low-phosphorus coatings in the plane (1 1 1). A similar trend was observed for the diffraction plane (2 0 0).

On the other hand, in the case of the heat-treated Ni–P coatings, the Ni_3P crystallite size increased with the increasing P content, as seen in Figure 6. The Ni_3P crystallite size dependence on the P content was studied for diffractions with the highest intensity. The most distinctive diffraction angles were $2\theta \approx 41.7°$, $42.8°$, $43.6°$, $45.3°$, $46.6°$, and $52.7°$, which corresponds to the diffraction planes (3 2 1), (3 3 0), (1 1 2), (4 2 0), (1 4 1), and (3 1 2), respectively. Increasing crystallite size of Ni_3P is related to increasing P content. With a higher P content, a Ni_3P phase fraction is formed and combined to form coarser particles. Meanwhile, in the case of low phosphorus coatings, the formed Ni_3P phase is in the form of fine-grained precipitates distributed in the Ni–P matrix.

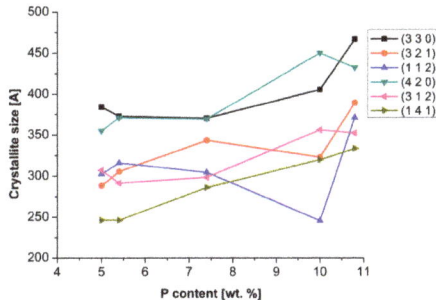

Figure 6. The effect of Ni_3P crystallite size on the P content in heat-treated Ni–P coatings.

Based on the literature [10,14], it is evident that the chemical composition and the size of the Ni_3P phase can affect the resulting microstructure and hardness of the Ni–P coatings.

Higgs [18] states that the heat-treated coatings showed fine-grained intermetallic precipitates of Ni_3P in the Ni–P matrix. The author also pointed out that the size of Ni_3P precipitates in the Ni–P matrix depended on the temperature. According to our study, the size of Ni_3P precipitates is also related to the P content (Figure 6).

Finer precipitates of Ni_3P may be responsible for the increased hardness or the improvement of other mechanical properties [14,25]. With reference to measured results of microhardness of the Ni–P coatings (Figure 3) and their microstructural characteristics (Figures 5 and 6), it is evident that the microhardness depends not only on the P content, but also on the size and distribution of Ni_3P

precipitates in the case of heat-treated coatings. Therefore, the heat treatment leads to a significant increase of microhardness of the Ni–P coatings.

However, the heat treatment also affects the substrate. The temperature of 400 °C influences the structure of Mg alloys more than in the case of steels. The heat treatment of the AZ91 alloy led to the dissolution of the discontinuous precipitate $Mg_{17}Al_{12}$. The difference in the thermal expansion coefficient of AZ91 Mg alloy and the Ni–P coating may lead to tension at the interface. However, there was no observable impact (visible cracks or delamination) at the Mg substrate/Ni–P coating interface.

4. Conclusions

The electroless Ni–P coatings with various P content were deposited on AZ91 Mg alloys and subsequently heat-treated at 400 °C for 1 h.

As-deposited Ni–P coatings showed the decrease in microhardness with increasing P content. Heat-treated Ni–P coatings showed a similar trend. However, the heat-treated coatings reached significantly higher microhardness values.

From the XRD analysis, it was determined that the microstructure of the as-deposited high-phosphorus coatings was amorphous, and with the decreasing P content they become more crystalline. Heat-treated Ni–P coatings were completely crystalline, and a presence of crystalline Ni and the intermediate Ni_3P phase in the coating was observed. It was observed that the Ni crystallite size in the coating decreased with the increasing P, both for as-deposited and heat-treated Ni–P coatings. On the other hand, the crystallite size of Ni_3P increased with increasing P content in the coating.

In terms of the precipitation hardening process, the heat-treated Ni–P coatings reached higher microhardness values than the as-deposited coatings. This is due to the presence of a large number of intermetallic precipitates of Ni_3P.

The influence of temperature during the heat treatment led to the dissolution of the discontinuous precipitate $Mg_{17}Al_{12}$. Despite the substrate microstructural changes and difference in thermal expansion coefficients, this did not lead to the delamination or visible cracking of the coating. The prepared coatings were uniform and with no visible defects.

Author Contributions: Conceptualization, M.B. and M.K.; methodology M.B., J.M. and J.W.; validation, M.B., M.K. and J.W.; formal analysis, M.B. and M.K.; investigation, M.B. and M.K.; resources, M.B. and J.W.; data curation, M.B. and J.M.; writing—original draft preparation, M.B.; writing—review and editing, M.B. and M.K. visualization, M.B. and M.K.; supervision, J.W.; project administration, J.W.; funding acquisition, J.W.

Funding: This work was supported by project Nr. LO1211, Materials Research Centre at FCH BUT- Sustainability and Development (National Program for Sustainability I, Ministry of Education, Youth and Sports).

Conflicts of Interest: The authors declare no conflict of interest.

References

1. Friedrich, H.; Mordike, B.L. *Magnesium Technology: Metallurgy, Design Data, Applications*; Springer: Berlin, Germany, 2006.
2. Czerwinski, F. *Magnesium Alloys: Design, Processing and Properties*; InTech: Rijeka, Croatia, 2011.
3. Buchtík, M.; Kosár, P.; Wasserbauer, J.; Tkacz, J.; Doležal, P. Characterization of electroless Ni–P coating prepared on a wrought ZE10 magnesium alloy. *Coatings* **2018**, *8*, 96. [CrossRef]
4. Ambat, R.; Zhou, W. Electroless nickel-plating on AZ91D magnesium alloy: Effect of substrate microstructure and plating parameters. *Surf. Coat. Technol.* **2004**, *179*, 124–134. [CrossRef]
5. Gray, J.E.; Luan, B. Protective coatings on magnesium and its alloys—A critical review. *J. Alloy. Compd.* **2006**, *336*, 88–113. [CrossRef]
6. Liu, Z.; Gao, W. The effect of substrate on the electroless nickel plating of Mg and Mg alloys. *Surf. Coat. Technol.* **2006**, *200*, 3553–3560. [CrossRef]
7. Tkacz, J.; Minda, J.; Fintová, S.; Wasserbauer, J. Comparison of electrochemical methods for the evaluation of cast AZ91 magnesium alloy. *Materials* **2016**, *9*, 925. [CrossRef] [PubMed]

8. Seifzadeh, D.; Mohsenabadi, H.K. Corrosion protection of AM60B magnesium alloy by application of electroless nickel coating via a new chrome-free pretreatment. *Bull. Mater. Sci.* **2017**, *40*, 407–415. [CrossRef]
9. Mallory, G.O.; Hajdu, J.B. *Electroless Coating: Fundamentals and Applications*; Noyes Publishing, William Andrew: Norwich, NY, USA, 2009.
10. Riedel, W. *Electroless Nickel Plating*; ASM International: Metals Park, OH, USA, 1991.
11. Duncan, R.N. The metallurgical structure of electroless nickel deposits: Effect on coating properties. *Plat. Surf. Finish.* **1996**, *8*, 65–69.
12. Agarwala, R.C.; Agarwala, V. Electroless alloy/composite coatings: A review. *Sadhana* **2003**, *3–4*, 475–493. [CrossRef]
13. Parkinson, R. *Properties and Applications of Electroless Nickel*; Nickel Development Institute: Toronto, ON, Canada, 1997.
14. Keong, K.G.; Sha, W.; Malinov, S. Hardness evolution of electroless nickel–phosphorus deposits with thermal processing. *Surf. Coat. Technol.* **2003**, *168*, 263–274. [CrossRef]
15. Sudagar, J.; Lian, J.; Sha, W. Electroless nickel, alloy, composite and nano coatings—A critical review. *J. Alloy. Compd.* **2013**, *571*, 183–204. [CrossRef]
16. Sampath Kumar, P.; Kesavan Nair, P. Studies on crystallization of electroless Ni–P deposits. *J. Mater. Process. Technol.* **1996**, *56*, 511–520. [CrossRef]
17. Niksefat, V.; Ghorbani, M. Mechanical and electrochemical properties of ultrasonic-assisted electroless deposition of Ni–B–TiO$_2$ composite coatings. *J. Alloy. Compd.* **2015**, *633*, 127–136. [CrossRef]
18. Higgs, C.E. The effect of heat treatment on the structure and hardness of an electrolessly deposited nickel–phosphorus alloy. *Electrodepos. Surf. Treat.* **1974**, *2*, 315–326. [CrossRef]
19. Ashtiani, A.; Faraji, A.S.; Amjad Iranaghi, S.; Faraji, A.H. The study of electroless Ni–P alloys with different complexing agents on Ck45 steel substrate. *Arab. J. Chem.* **2017**, *10*, 1541–1545. [CrossRef]
20. Gu, C.; Lian, J.; Li, G.; Niu, L.; Jiang, Z. Electroless Ni–P plating on AZ91D magnesium alloy from a sulfate solution. *J. Alloy. Compd.* **2005**, *391*, 104–109. [CrossRef]
21. Hu, B.; Sun, R.; Yu, G.; Liu, L.; Xie, Z.; He, X.; Zhang, X. Effect of bath pH and stabilizer on electroless nickel plating of magnesium alloys. *Surf. Coat. Technol.* **2013**, *228*, 84–91. [CrossRef]
22. Kundu, S.; Das, S.K.; Sahoo, P. Properties of electroless nickel at elevated temperature—A review. *Procedia Eng.* **2014**, *97*, 1698–1706. [CrossRef]
23. Mainier, F.B.; Fonseca, M.P.C.; Tavares, S.S.M.; Pardal, J.M. Quality of electroless Ni–P (Nickel–Phosphorus) coatings applied in oil production equipment with salinity. *J. Mater. Sci. Chem. Eng.* **2013**, *1*, 1–8. [CrossRef]
24. Keong, K.G.; Sha, W.; Malinov, S. Crystallisation kinetics and phase transformation behaviour of electroless nickel—Phosphorus deposits with high phosphorus content. *J. Alloy. Compd.* **2002**, *334*, 192–199. [CrossRef]
25. Guo, Z.; Keong, K.G.; Sha, W. Crystallisation and phase transformation behaviour of electroless nickel phosphorus platings during continuous heating. *J. Alloy. Compd.* **2003**, *358*, 112–119. [CrossRef]

© 2019 by the authors. Licensee MDPI, Basel, Switzerland. This article is an open access article distributed under the terms and conditions of the Creative Commons Attribution (CC BY) license (http://creativecommons.org/licenses/by/4.0/).

Article

Surface Activation and Characterization of Aluminum Alloys for Brazing Optimization

Sara Ferraris *, Sergio Perero and Graziano Ubertalli

Department of Applied Science and Technology, Politecnico di Torino, 10129 Torino (TO), Italy
* Correspondence: sara.ferraris@polito.it; Tel.:+39-1-1090-5768

Received: 17 June 2019; Accepted: 19 July 2019; Published: 23 July 2019

Abstract: Brazing of Al-alloys is of interest in many application fields (e.g., mechanical and automotive). The surface preparation of substrates and the in depth investigation of the interface reaction between aluminum substrates and brazing materials is fundamental for a proper understanding of the process and for its optimization. The interaction between two aluminum based substrates (Al5182 and Al6016) and two studied brazing materials (pure Zn and for the first time ZAMA alloy) has been studied in simulated brazing condition in order to define the best surface preparation conditions and combination substrate-brazing material to be used in real joining experiments. Three different surface preparations were considered: polishing and cleaning, application of flux and vacuum plasma etching (Ar) followed by sputtering coating with Zn. Macroscopic observation of the samples surface after "brazing", optical microscopy, and microhardness measurements on the cross-section and XRD measurements on the top surface gave a comprehensive description of the phenomena occurring at the interface between the substrate and the brazing alloy which are of interest to understand the brazing process and for the detection of the best conditions to be used in brazing. Plasma etching (Ar) followed by sputtering coating with Zn resulted a promising solution in case of Al5182 brazed with Zn, while the addition of flux was more effective in case of Al6016 substrate. ZAMA alloy demonstrated good interface reactivity with both Al6016 and Al5182 alloys, particularly on only cleaned surfaces.

Keywords: aluminum alloys; brazing; surface preparation; interface reactions; joining; microstructure; phase/composition in reaction layer

1. Introduction

Joining of Al alloys is a crucial point in the realization of components for mechanical, automotive and aerospace industries. Brazing of Al alloys with Zn-based filler materials is a promising joining solution for its high chemical affinity for aluminum, moderate brazing temperature, good corrosion resistance and mechanical properties, as well as for the obtainment of components, which can be used at high temperature and can be recycled at the end of their life (in contrast to adhesive joints) [1,2].

AlSiZnSrTi [3], Zn-Al (Al 2%–12%) [4], and Zn-14Al (hypereutectic) [1] alloys were successfully used for the brazing of Al-alloys plates. Moreover, 6.2 wt % Al, 4.3 wt % Cu, 1.2 wt %, Mg, 0.8 wt % Mn, 0.5 wt % Ag and balance Zn [5,6], pure Zn and Zn-2%Al [7], and Al-Cu-Mg and Al-Si-Mg-17%Ti metal glasses [8] were employed for the realization of Al-alloy dense sheets–Al-foam sandwiches with good results.

One of the main obstacles to the effective brazing of Al alloys is the presence of a stable oxide layer on the surface with high melting temperature and poor reactivity, which act as a barrier for metallurgical interaction between the aluminum alloy substrate and the brazing material [9,10]. Particularly critical from this point of view are the Al alloys of the 5000 series because of high magnesium content, which promotes the development of a thick and stable magnesium oxide layer.

A fist way to remove the surface oxide and improve brazing ability of Al alloys is the mechanical removal of the first surface layer by abrasive papers; this operation also increases surface roughness and consequently the mechanical interlocking between substrate and brazing material [7,11,12]. On the other hand, by this method, surface reoxidation is not hampered and the oxide layer can develop again on the surface, especially at high temperature during brazing. In order to overcome such problem, the use of fluxes as deoxidizing and protective agents has been widely employed [7]. In particular, Cs containing fluxes and the addition of activating agents (e.g., $ZnCl_2$) have been suggested for Al alloys with high Mg content for the effective removal of MgO and for the surface preservation from reoxidation [3,4,12,13].

Plasma etching with Ar followed by the sputtering deposition of a Cu layer (10 μm) has also been suggested for the removal of the oxide layer (plasma etching), protection from reoxidation (coating immediately after etching under vacuum), and deposition of a layer (Cu), which can be employed as brazing material [9].

Finally, a zincate treatment (sequential soaking in 1.2 M NaOH, 69.5% HNO_3 and commercial zincate solution) has been proposed for the improvement of bonding between aluminum and Al-foams [14].

In the present research work specimens of Al6016 and Al5182 alloys have been considered as Al substrates. Pure zinc and, for the first time, ZAMA alloy have been studied as possible brazing materials. Pure zinc has been selected on the basis of the previously obtained good results in Al-alloy-Al-foam brazing [7], while ZAMA alloy has been tested because Cu has a high solubility in Al and it can improve mechanical and corrosion resistance of the joint. Moreover, the presence of Cu and Mg in the metallic glasses used in previous brazing experiment with Al foams [8] led to good results. Three different preparations of the substrates have been explored: oxide removal by abrasive paper and ultrasonic cleaning, application of a flux, and vacuum plasma Ar etching followed by sputtering deposition of a Zn layer. The interaction between the substrate and the brazing material has been investigated in simulated brazing condition by means of macroscopic observations, optical microscopy, and microhardness measurements on the cross-section and X-ray diffraction (XRD) measurements on the top surface.

An in depth investigation of the phenomena occurring at the interface between the substrate and the brazing material is crucial for the understanding of the brazing process and for its optimization.

The study identified the most promising solutions, in terms of surface preparation and brazing materials, to be used in further real brazing experiments (e.g., joining of aluminum plates to aluminum foams). Moreover, from this research a protocol for the characterization of brazed joints and interfaces has been developed and proposed.

2. Materials and Methods

Specimens of Al6016 and Al5182 alloys were taken off from 1 mm thick plates (compositions are given in Table 1), and were used as brazing substrates since they are of interest in many industrial applications, such as automotive components for Body in white (BIW).

Table 1. Chemical composition of the Al alloys used as substrates [15].

Alloy	Elements (wt %)								
	Si	Fe	Cu	Mn	Mg	Cr	Zn	Ti	Al
Al6016	1.00–1.50	0.50	0.20	0.20	0.26–0.60	0.10	0.20	0.15	rem
Al5182	0.20	0.35	0.15	0.20–0.50	4.00–5.00	0.10	0.25	0.10	rem

Pure zinc (Zn, Lucas Milhaupt™, Cudahy, WI, USA.) and ZAMA (Al 3.9%–4.3%, Cu 0.75%–1.25%; Mg 0.03%–0.06%; Fe 0.05%; Pb 0.005%; Sn 0.002%; Cd 0.005%; Zn rem; Dynacast Italia SpA, Grosso-TO, Italy) alloy were used as brazing alloys. Zn was selected for its high chemical affinity for Al and for the good results obtained in the past for the brazing of Al6016 to Al foams by the authors [7]. ZAMA was tested for the first time in the present research work as possible brazing material.

Different surface preparations have been considered in order to investigate surface reactions and possible optimization strategies for brazing. In fact, aluminum alloys spontaneously passivate by a surface oxide layer which could hamper surface reactivity and brazing effectiveness because of its inertness and high melting temperature. In order to favor the brazing process, the surface oxide layer should be removed.

The surface of all the specimens was activated with mechanical oxide removing by abrasive paper grinding (320 grit), and cleaned by ultrasonic washing in ethanol (10 min, 60 °C). This was the preparation of the first set of specimens (1).

In the second set of specimens the brazing was effected with the addition of a cesium fluoroaluminate containing flux (FLUX-AL6, Stella Welding, Albizzate (VA), Italy).

In the third one, a surface of each specimen was vacuum plasma Ar etched (15 min, 200 W in radio frequency (RF), operating pressure 1 Pa. The instrument used was a Kenosistec™ (Binasco, Italy)sputtering equipped with three confocal three inches cathode), then Zn coated by sputtering (30 min, 100 W in direct current (DC), operating pressure 6×10^{-1} Pa), with the same instrument.

The application of flux is adopted for a more effective surface oxide layer removal and to avoid its restoration, as described in [3,4,7,13]. Analogously, Ar-plasma etching guarantees a complete surface cleaning and deoxidation as it has been reported by Hu et al. [9]. In this research, plasma etching has been coupled with the deposition (in the same atmosphere and reaction chamber) of a Zn layer, with the aim to facilitate the interaction with Zn-based brazing alloys and avoid surface reoxidation.

In this research the brazing process conditions have been simulated (atmosphere, time, temperature) and the samples considered for the test were constituted only from the substrate and the brazing material, in the form of strips or flatten chips, placed on the top surface of the substrate (~0.16 g of brazing alloys were considered as suggested by Dai et al. [4])

The brazing process was conducted in a tubular furnace (Carbolite, Hope Valley, UK) in an Ar atmosphere with a heating rate of 10 °C/min and a dwell time of 5 min. The Ar atmosphere was chosen to prevent high temperature surface oxidation phenomena. The temperature of 480 °C has been selected for brazing experiments with pure zinc, while the temperature of 520 °C has been used for experiments with ZAMA. Preliminary reduced set of experiments were conducted to evaluate the best temperature and time for the chosen alloys.

Surface reactions between selected substrates and brazing alloys were investigated by means of spreading tests, as described by numerous authors [4,12,16–18]: The samples were macroscopically observed in order to determine the shape of the brazing alloy after treatment and to determine (in a semiquantitative way) the entity of spreading. Moreover, the samples were transversally cut and one half of each sample metallographically prepared to analyze the cross-section at the optical microscope (Reichert-Jung MeF3, Leica Microsystems Srl, Buccinasco (MI), Italy) and to investigate the brazing alloy–substrate interface without etching. The Vickers micro hardness (10 g load—Remet HX 1000, Remet, Casalecchio di Reno (BO), Italy) was measured in correspondence of the brazing alloy–substrate interface (reaction layer) and in the bulk substrate material. X-ray diffraction ((XRD) Rigaku D-MAX3, Rigaku Europe SE, Neu-Isenburg, Germany) measurements were performed on the frontal surface of the second half of each sample in order to detect the phases present on the surface after the simulating brazing process.

Finally, the cross-section of the most promising samples was analyzed by means of field emission scanning electron microscopy equipped with field-emission scanning electron microscopy–energy-dispersive spectroscopy (FESEM-EDS, SUPRA™ 40, Zeiss, Berlin, Germany).

The setup of the spreading test, developed in the present research, and the combination of different analytical techniques for the characterization of the top surface and the cross-section of the prepared samples have been demonstrated an effective procedure in order to understand the interface reactions between Al substrates and brazing materials. This investigation can be employed for the design and realization of promising brazing experiments.

A scheme of the test procedure and rationale is reported in Figure 1.

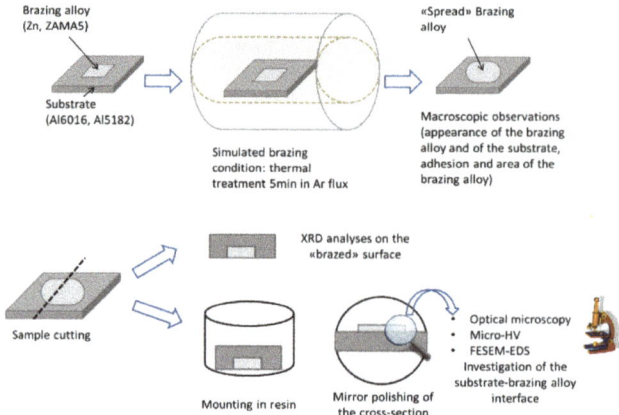

Figure 1. Scheme of the research setup and rationale.

3. Results

3.1. Macroscopic Appearance of Samples before and after Simulating Brazing

The comparison of the visual appearance of the top surfaces before and after the spreading test allows to observe the shape of the brazing alloy after treatment and to determine (in a semiquantitative way) the entity of spreading. This is a first indication of the reactions occurred during the treatment. Moreover the qualitative estimation of the adhesion strength between the substrates and the brazing materials after the test can be helpful for the determination of the stability of the produced reaction layers. Specific comments on the obtained results are reported below. The macroscopic appearance of samples before and after brazing is shown in Figure 2.

Even if it is not possible to quantify the variation in the brazing material area, some preliminary considerations can be made from the images in Figure 2:

- The alteration in the brazing alloy piece shape suggests the formation of a certain amount of liquid during the spreading testing in all the tested conditions.
- Depending on the amount of the formed liquid and on high temperature wettability of the brazing alloy on the substrate, a different shape of the solidified brazing alloy can be observed.
- In the cases of Al5182 substrate treated with flux and ZAMA brazing alloy and for Al6016 substrate treated with flux with both Zn and ZAMA brazing alloys, the brazing material assumed a drop-like shape after the treatment.
- In the case of Al5182 and Al6016 treated with flux, the ZAMA brazing stripes assumed a rounded particle shape after the test. The formation of these ZAMA particles came from the formation of liquid drops during the experiment which does not wet the substrate (poor wettability of the substrate for the liquid ZAMA at the selected temperature). This phenomenon can be associated with the behavior of flux at the selected temperature, which induces poor wettability for liquid ZAMA.

- For all the other substrate-brazing alloy combinations, the brazing materials maintain the initial strip shape with some spreading.

Moreover some differences can be evidenced in the adhesion strength of the brazing materials on the substrates:

- Zinc on just-cleaned Al6016 and on cleaned and flux-covered Al5182 substrates is easily removed after the treatment without evident signs on the substrates.
- Both Zinc and ZAMA alloy are removed from Zn-sputtered substrates, however leaving important tracks on the substrates.
- A good adhesion of the brazing material on the substrates have been observed, after the thermal treatment, for Zinc on Al6016 alloy treated with flux and ZAMA alloy on Al6016 just-cleaned and treated with flux as well as on Al5182 just-cleaned.

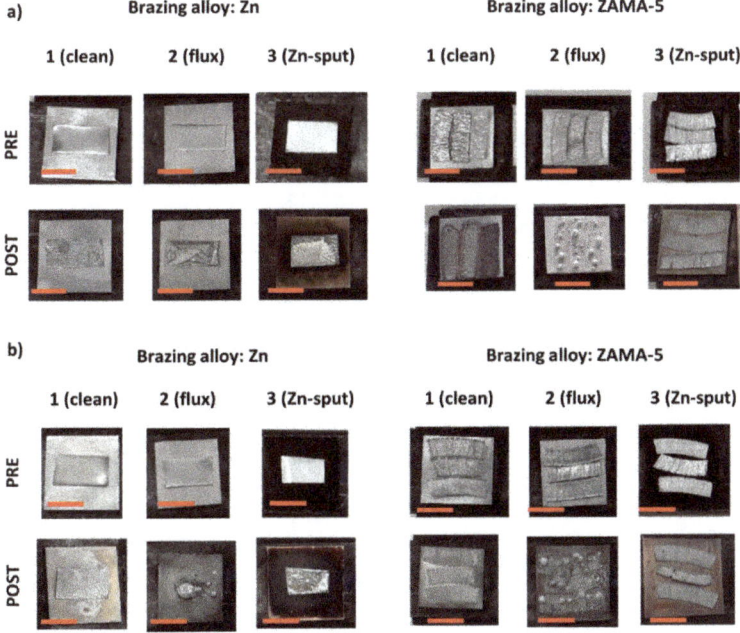

Figure 2. Macroscopic appearance of samples before and after brazing. (**a**) Al5182 substrate and (**b**) Al6016 substrate. Scale bar: 1 cm.

3.2. Metallographic Observations of the Cross-Sections

The observation at the optical microscope of the samples cross-section after metallographic preparation (resin mounting, mirror polishing, and no etching) is a useful strategy to investigate the thickness and the microstructure of the reaction layers and the quality of the interface produced between the substrate and the brazing material during the treatment. The optical microscope images of the transverse section of the samples are reported in Figure 3. The main observations are listed below.

- No evident reaction layers can be observed in case of the simulating brazing tests with Pure Zinc on both just-cleaned Al6016 and Al5182 alloys.
- A moderate reaction layer (less than 100 μm) can be noted for the pure Zinc and the ZAMA on Zn-sputtered Al6016 substrates.

- In the cases of pure Zinc and ZAMA on Al5182 alloy substrate treated with flux, the brazing alloy formed drops without a continuous interface with the substrate and without evident reaction layer. A discontinuous reaction zone with evident melt and solidified brazing alloy pieces can be documented for ZAMA on Al6016 treated with flux.
- Zinc on Al6016 treated with flux forms spread drops with a continuous interface with the substrate and an evident reaction layer (hundreds of microns).
- Important reaction can be documented also for Zinc on Zn-sputtered Al5182 substrate and for ZAMA on just-cleaned Al5182 substrate; in these samples the reaction layer is homogeneous on the whole surface (hundreds of microns in thickness) and presents the typical appearance of melt and resolidified metal with evident grains, well-visible because of the remarkable amount of porosity, probably connected with liquid shrinkage.
- Finally a continuous, but thinner than on Al5182 (less than 100 µm), reaction layer can be observed for ZAMA on just-cleaned Al6016 sample.

These results are in accordance with the previously reported macroscopic observations.

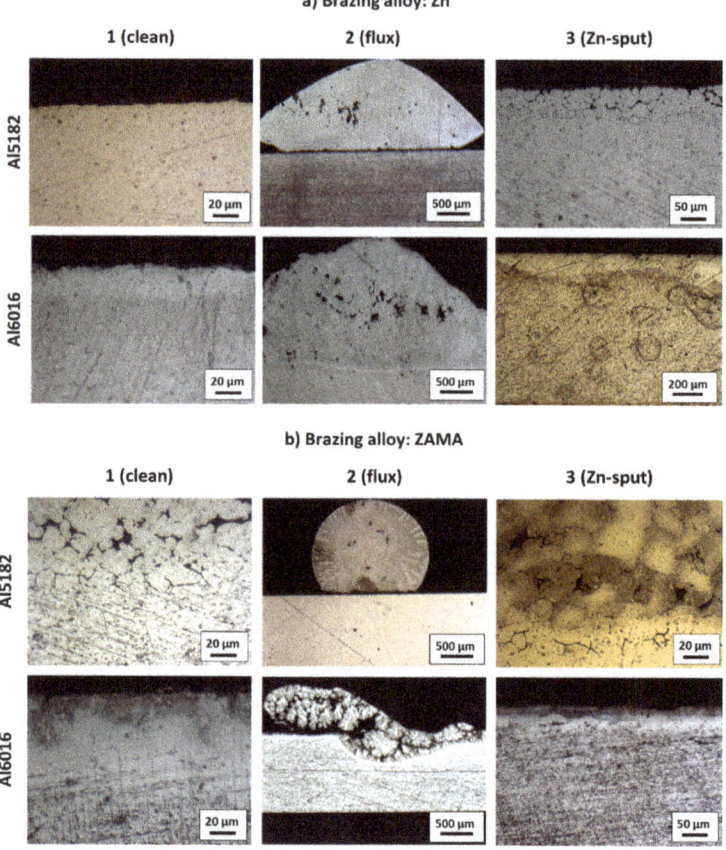

Figure 3. Optical microscope images of the transverse section of the samples (no etching): (**a**) Zn as brazing material; (**b**) ZAMA as brazing material.

3.3. Microhardness Measurements on the Cross-Sections

Microhardness measurements, performed on the most representative zones of the metallographic transverse samples after brazing, are reported in Table 2. These values give information on the hardness properties of the microconstituents in the melted zones, on reaction layers and on the base materials after the spreading tests.

Table 2. Microhardness measurements.

Substrate	HV0.01					
	Brazing Alloy: Pure Zn			Brazing Alloy: ZAMA5		
	1 (clean)	2 (flux)	3 (Zn sput)	1 (clean)	2 (flux)	3 (Zn sput)
Al5182	Surf. 39 Bulk 41	Drop 35 Bulk 43	Surf. 62 Bulk 42	Surf. 68 Interf. 55 Bulk 41	Drop 51 Bulk 37	Surf. 78 Interf. 78 Bulk 42
Al6016	Surf. 46 Bulk 43	Drop 72 Interf. 73 Bulk 44	Surf. 39 Bulk 40	Surf. 64 Interf. 68 Bulk 44	Surf. 73 Interf. 79 Bulk 44	Surf. 44 Interf. 33 Bulk 43

The microhardness values of the bulk materials (after brazing) are between 37 and 44. A significant increase in the surface hardness can be observed for Al5182 Zn-sputtered and Al6016 treated with flux after the use of pure Zn as brazing alloy (reaching values of 62 and 72, respectively) and for almost all the considered surfaces after the use of ZAMA as brazing alloy (values ranging from 64 for clean Al6016 to 78 for Zn-sputtered Al5182).

Also in this case the results confirm macroscopic and optical observations; in presence of significant reactions between the brazing alloy and the substrate, an important increase of hardness in the reaction surface was obtained. The increase in microhardness in the reaction layer, compared to the bulk Al-alloy substrate, can be attributed to the contribution of Zn to the Al solid solution and to second phase formation in the reaction layer (observed at the optical microscope). The effect of zinc oxides (observed only on the outer surface of some samples by means of XRD measurements) can be considered negligible on hardness measurements.

3.4. XRD Measurements on the Top Surface

XRD measurements on the top surfaces of samples after the simulated brazing treatment were performed in order to investigate the possible presence of intermediate phases, formed during the reaction between the substrates and the brazing materials through the detection of specific crystallographic structures on a thin layer of the surface (micron).

The XRD measurements (Figures 4 and 5) show that the Al5182 and Al6016 specimens with just-cleaned surfaces, the Al5182 once treated with flux and Al6016 sputtered with Zn present Al as the main phase on the surface after the spreading test, with negligible contribution of Zn. An increase in the Zn surface content (with the consequent decrease intensity of Al peaks) can be observed for Al6016 specimens treated with flux and for Zn-sputtered Al5182 once after the spreading test with Zn and for all the substrates after the spreading test with ZAMA. Moreover Zn-sputtered Al5182 samples after the spreading test with Zn and Al5182 just-cleaned or Zn sputtered after the spreading test with ZAMA highlight the presence of intermediate phases of binary and ternary phase diagrams such as Al40Zn60, Al70Zn30, Cu1.44MgSi0.56, Al25Mg37.5Zn37.5, and Mg32(Al, Zn)49.

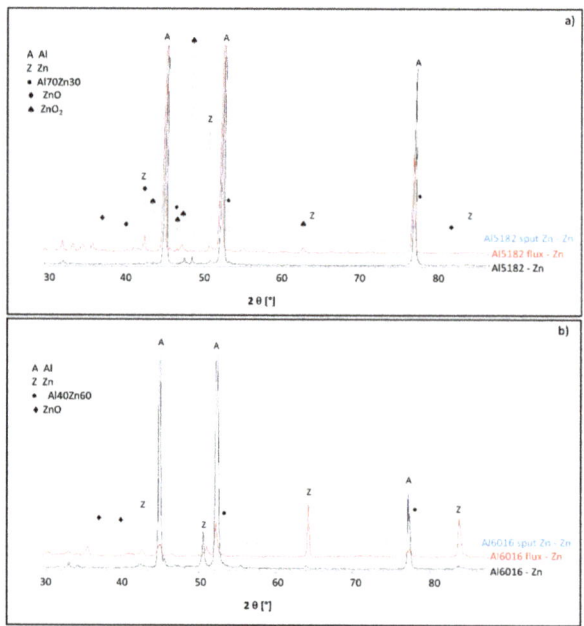

Figure 4. XRD spectra of the top surface of samples after the "spreading test". (Zn as brazing material.): (**a**) Al 5182 substrate; (**b**) Al 6016 substrate.

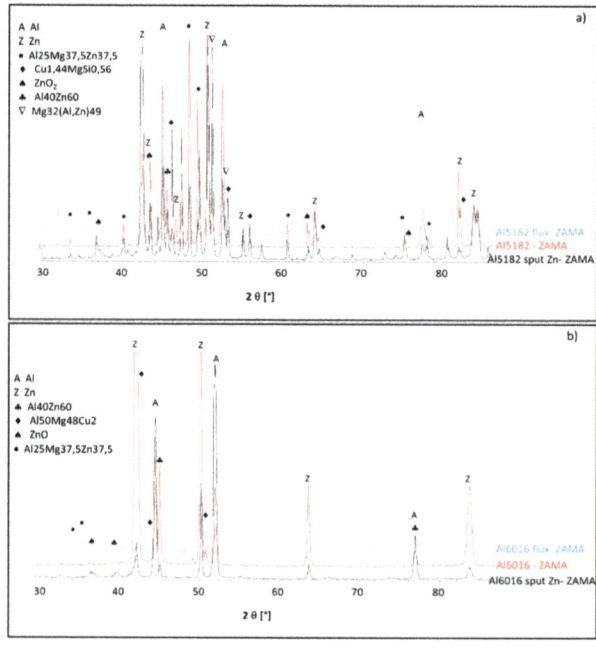

Figure 5. XRD spectra of the top surface of samples after the "spreading test". (ZAMA as brazing material.): (**a**) Al 5182 substrate; (**b**) Al 6016 substrate.

Minor peaks can be observed in some of the reported diffractograms but in the present study their influence can be considered negligible to understand the mechanism of reaction and formation of reaction layers. They are only present in a very thin surface layer (micron).

These results can be integrated with the microprobe analysis to better define the phases formed during the spreading tests (Section 3.5).

The detection of Zn rich phases, by means of XRD analyses on the top surface of the samples, in presence of significant reaction layers (hundreds of microns) observed at the optical microscope on the transverse sections, confirms the thickness of the reactions layers in accordance to XRD penetration depth.

XRD spectra of samples after the "spreading" test are reported in Figure 4 (Zn as brazing material) and Figure 5 (ZAMA as brazing material).

3.5. FESEM-EDS Analyses

FESEM-EDS analyses on the cross-section of the samples have been performed on the most promising samples in order to obtain indications of the chemical elements and of their rates in specific areas of the reaction layer and interfaces and support the formation of specific intermediate phases (hypothesized from XRD analyses, reported above).

The transverse section of samples, evidencing good brazing effect, was observed by means of FESEM-EDS analyses. FESEM cross-sectional observations and EDS analyses of selected areas for Al6016 flux-Zn (Figure 6), Al5182-Zn-sput-Zn (Figure 7), Al6016-ZAMA (Figure 8), and Al5182-ZAMA (Figure 9) are reported.

In the simulated brazing process of the Al6016 alloy with Zn using flux, liquid drops were formed that did not spread on the surface but kept the rounded shape. However, the liquid reacted locally with the aluminium plate causing its partial melting. The microstructure resulting from solidification of the liquid was of dendritic type, especially in the part of the drop outside the plate, where the heat removal rates were lower (Figure 6a). The metallographic observations carried out at the interface between the molten zone and the bulk material of the plate showed a biphasic morphology in the material solidified by the liquid (Figure 6b) and the microanalytical investigations (Figure 6c) confirmed the presence of phase β'(Al70Zn30) rich in aluminium (areas 1 and 3 of investigation) and the phase rich in zinc (area 2 of investigation). At the interface between the pre-existing drop and the aluminium plate the microanalytical results show a much richer composition in aluminium, with a higher temperature of existence of the liquid, then solidified, but compatible with the chosen brazing treatment temperatures.

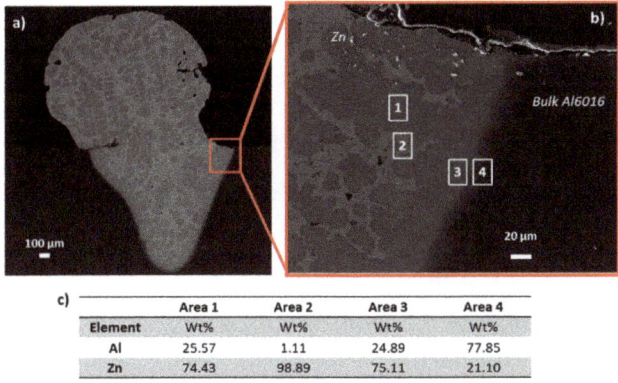

Figure 6. Field-emission scanning electron microscopy–energy-dispersive spectroscopy (FESEM-EDS) analyses of Al6016-flux-Zn. (**a**) Micrograph of the drop/reaction layer; (**b**) magnification of the reaction layer/interface; and (**c**) semiquantitative EDS analyses of selected area evidenced in panel (**b**).

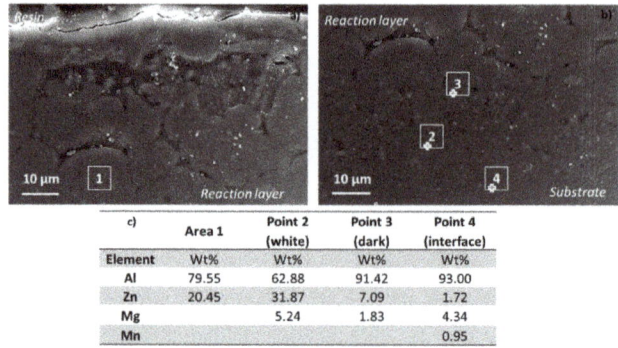

Figure 7. FESEM-EDS analyses of Al5182 Zn-sput-Zn. (**a**) Micrograph of the reaction layer; (**b**) magnification of the reaction layer/interface; and (**c**) semiquantitative EDS analyses of selected areas evidenced in panel (**b**).

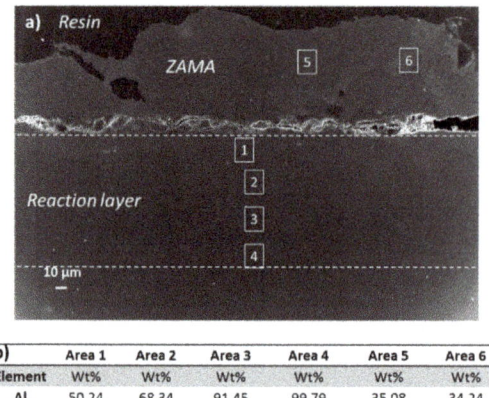

Figure 8. FESEM-EDS analyses of Al6016-ZAMA. (**a**) Micrograph of the reaction layer and (**b**) semiquantitative EDS analyses of selected areas evidenced in panel (**a**).

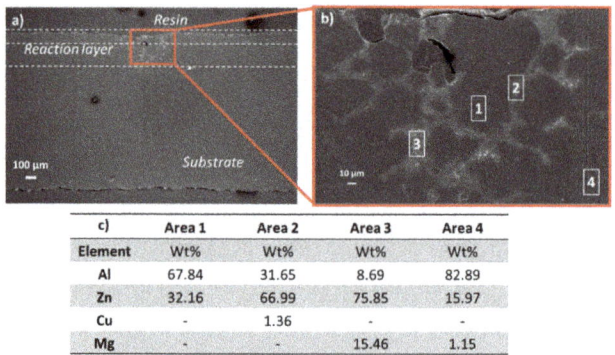

Figure 9. FESEM-EDS analyses of Al5182-ZAMA. (**a**) Low magnification micrograph of transverse section reaction/layer substrate interface; (**b**) high magnification of the reaction zone; and (**c**) semiquantitative EDS analyses of selected areas evidenced in panel (**b**).

It should also be noted that the biphasic dendritic microstructure observable in the solidified zone during cooling after brazing is markedly extended in the drop zone, above the level of the plate surface. This means that the molten zinc has reacted with the plate, enriching itself in aluminium, and therefore decreasing the temperature of existence of the liquid mixture, in accordance with the Al-Zn binary state diagram.

As far as Al5182 Zn-sput-Zn sample is concerned, the images shown refer to the surface layer and the first reaction layer (Figure 7a) passage between reaction layer and matrix (Figure 7b).

In this sample an homogeneous superficial reaction layer (without drop formation) is observed in which biphasic morphologies with grains of metal matrix of the plate are detectable by optical and electronic microscopy separated from eutectic mixtures whose quantity progressively decreases from the outside towards the inside in the reaction layer.

Point 1 evidences partially melted aluminium grains and mixed with Zn after solidification, point 2 links β' with composition (70% Al and 30% Zn), point 3 shows grains of aluminium with only incipient fusion and superficially enriched with Zn, and point 4 presents only traces of zinc diffusion.

The sample of the spreading test of the Al alloy 6016 simulating brazing with ZAMA alloy (Figure 8) showed the formation of a liquid phase which showed a certain spreading on the surface. The liquid reacted with the aluminium plate, enriched with aluminium itself and, after solidification, as confirmed by the EDS analyses on the ZAMA strip after brazing (areas 5 and 6 in Figure 8) which are in accordance to the Al-Zn phase diagram. The interface between the brazing alloy and the surface of the plate shows marked longitudinal cracks that have compromised the relative adhesiveness and soundness of the joint. Below these cracks the reaction and diffusion layer of the ZAMA brazing alloy with the aluminium plate can be identified. While at the microstructural level there are no detectable grain boundaries or microstructures, at compositional level the microprobe investigations, carried out on the reduced areas (1–4), show a progressive decrease of zinc and a consequent increase in aluminium content going towards the center of the plate, for a thickness of ~110 μm.

The sample of the spreading test of the Al alloy 5182 simulating brazing with ZAMA alloy (Figure 9) shows a continuous surface layer or reaction 100 micrometers thick and a zone, evidenced in Figure 9a) with a deeper reaction. This region at higher magnification (Figure 9b) shows a morphology produced from solidification with grains surrounded from a fine eutectic microstructure. The microanalyses conducted in the single-phase microstructure, area 1, evidences an Al–Zn ratio of 70:30, while in the single-phase of area 4, at the boundary of the reaction zone, a certain amount of zinc diffuse in Al–Mg plate. In the eutectic microstructure a remarkable amount of zinc is detected (area of analysis 2) together with a high amount of magnesium (area 3 of analysis).

4. Discussion

The "spreading test" has been designed in order to investigate the interface reactions between two Al alloys substrates (Al5182 and Al6016) and two possible brazing materials (pure Zinc and ZAMA5 alloy) in simulated brazing conditions.

The test gave information on the interactions between the substrate and the brazing material in terms of: formation of brazing material drops and their shape/dimension, development of a diffusion layer at the interface and formation of a reaction layer on the substrate surface. The analyses of these information, by means of visual, optical, and electron microscopy observations and microhardness and XRD measurements allowed understanding of the reactions that occur between the brazing material and the substrate at brazing temperature and to individuate the most promising conditions for real brazing experiments.

In the present research several surface treatments (grinding and ultrasonic cleaning, application of flux or argon etching followed by Zn sputtering) were explored in order to remove the oxide layer (constituted by Al or Mg oxides) always present onto Al alloys with the aim to improve surface reactivity for brazing. Moreover simulated brazing experiments were conducted in Ar atmosphere in order to avoid significant reoxidation in temperature. Few nanometers of oxides can be formed in any case during the thermal treatment even in controlled atmosphere; however this fact is limited and occurs simultaneously to the liquid formation and the beginning of the substrate-brazing material interaction, therefore its effect can be considered negligible in the formation of the interface layer.

The results suggest that zinc oxides, sometime observed by means of XRD measurements, did not hamper the formation of the reaction layer.

Considering pure zinc as brazing material the most important reactions were obtained for Al6016 plates treated with flux, in accordance with previous researches of the authors [7].

Good surface reaction was also obtained for Zn-sputtered Al5182. This family (5000) showed the worst brazing properties, as reported in some papers [3,4,12,13], for the negative influence of the stable MgO formation in the reaction layer. The vacuum Ar etching effectively removed the Mg oxide layer, and the presence of the sputtered Zn layer avoids the reaction of the Mg of the plate with the oxygen (avoiding reoxidation), while it favors the reaction of the rich Zn liquid formed during the simulating brazing process. The Zn coating can therefore offer a promising solution in order to overcome the common difficulties in the brazing of 5xxx Al alloys.

Taking into account the ZAMA alloy strips as brazing material, it showed high reactivity with the Al substrate in the grinded or Zn sputtered conditions. This phenomenon can be explained considering the high solubility of Cu in Al and to the high reactivity of Cu and Mg towards Al, previously observed by the authors [8]. The most important reaction layers were obtained for just-cleaned surfaces and Zn-sputtered Al5182.

In these cases, reaction layers of hundreds of microns in thickness were reached, with Zn diffusion into the aluminum substrates (as demonstrated by optical microscopy observations, Figure 3, and FESEM-EDS analyses, Figures 8 and 9) and the formation of Zn-rich phases (as detected by XRD, Figure 5).

However, in this case the application of flux on the substrate did not improve the interaction between Al alloys and the brazing materials; in fact, sometimes it led to a more discontinuous reaction layer or hampered its formation.

Considering the comparison of the different characterization techniques, the main results and observations of the present research can be summarized as in the following.

In some cases, the macroscopic observation of the surface of the specimens (Figure 2) showed moderate variations in the sample appearance with an easy removal of the brazing strips from the substrate, an absence of significant reaction layer at the interface between substrate and brazing materials as also observed at the optical microscope (Figure 3) as well as a variation in the surface microhardness (Table 2) and a presence, almost exclusive, of aluminum in the XRD spectra (Figure 5). These are the cases of just-cleaned Al5182 and Al6016, Al5182 treated with flux and Zn-sputtered Al6016 in contact with pure Zn. These conditions can be considered unsuitable for successful brazing.

On the other hand, when macroscopic observations of the surface of the specimens (Figure 2) evidenced a significant alteration on the brazing strip shape (e.g., drop formation and strip depletion) and its stable anchorage to the substrate, very important reaction layers where observed at the optical microscope (Figure 3) together with a significant increase in surface microhardness (Table 2) and a development of Zn-rich phases on the specimen surface, detected at XRD (Figures 4 and 5). These are the cases of Zn sputtered Al5182 and flux treated Al6016 with pure Zinc and just-cleaned Al5182 and Al6016 and Zn sputtered Al5182 with ZAMA. These conditions can be considered suitable for successful brazing.

In order to sum up the interface reactions that drive the substrate-brazing material interaction in the here defined more promising conditions, the following explanations can be given. The mechanism beyond the effectiveness of Ar etching followed by Zn sputtering is based on the removal of the oxide layer (plasma etching) and the deposition of a protective layer (Zn sputtering) which avoid reoxidation and, in addition increase the compatibility and the reactivity for Zn in the brazing process, in accordance to what reported in the literature for Cu sputtered layers [9]. The use of flux act again supporting the removal of the oxide layer and hampering is formation during brazing [7]. ZAMA alloy has been used for the first time in the present research work as possible brazing material, in this case a high reactivity with just-cleaned Al alloys has been observed, according with the presence of Cu and Mg [8]. The use of flux was less effective with this material; this phenomenon can be correlated with the behavior of flux at the temperature used for spreading test with ZAMA, which can induce poor wettability for liquid ZAMA.

FESEM investigations highlight different possible morphological/compositional typologies of the reaction layers. For example, in the case of Zn on Al6016 treated with flux, the Zn strip formed a liquid drop which penetrated in the substrate for hundreds of microns, however the diffusion of zinc out of this drop interested a limited thickness (less than 50 μm), as shown in Figure 6. Differently, in the case of ZAMA on just-cleaned Al6016 (Figure 8), the ZAMA strip produced a liquid which solidified with different shape/dimension and produced detectable discontinuities. In this case the penetration of the liquid into the substrate is more limited but the diffusion of zinc is more deep (more than 100 μm). The presence of a mechanical and metallurgical continuity at the reaction layer can be of interest in order to obtain joints with good mechanical properties. On the other hand, the presence of some discontinuities can improve damping performances at the interface. The present research is focused on the investigation of the reactions between the substrates and the brazing materials and not on the development of a specific product, so at this level it is not possible to individuate a best behavior among the above described ones.

In summary plasma etching with Argon followed by Zn-sputtering resulted effective in the improvement of the reactivity between Al5182 substrates and pure zinc, offering a first promising solution for the brazing of this aluminum alloy. On the other hand, this procedure did not result so effective in the surface modification of Al6016. Moreover, ZAMA5 alloy has been explored for the first time as possible brazing alloy with encouraging results for both Al5182 and Al6016, especially without specific surface pre-treatments.

Both Ar plasma etching with Zn sputtering and the use of ZAMA alloy can represent interesting solutions to obtain good brazing results avoiding the use of flux, which in some case can be difficult to be removed and can increase brazing process times and cost [19].

5. Conclusions

The interaction between Al5182/Al6016 substrates with different surface preparations and Zn or ZAMA alloy has been investigated in simulated brazing conditions. The application of flux was the best strategy for the improvement of Al6016 reactivity with Zn; on the other hand, vacuum Ar plasma etching followed by Zn sputtering was necessary for the Al5182 substrate. The ZAMA alloy presented significant reactivity for both substrates without particular surface treatments, except cleaning. These conditions appear promising for the brazing of Al6016 and Al5182 alloys and will be explored in further research works.

Author Contributions: Conceptualization, S.F. and G.U.; Investigation, S.F., S.P. and G.U.; Data Curation, S.F., S.P. and G.U.; Writing—Original Draft Preparation, S.F., S.P. and G.U.; Writing—Review and Editing, S.F., S.P. and G.U.; Supervision, G.U.

Funding: This research received no external funding.

Acknowledgments: Dynacast Italia S.p.A., Grosso-TO, Italy, is kindly acknowledged for chips ZAMA providing.

Conflicts of Interest: The authors declare no conflict of interest.

References

1. Xiao, Y.; Ji, H.; Li, M.; Kim, J.; Kim, H. Microstructure and joint properties of ultrasonically brazed Al alloy joints using a Zn-Al hypereutectic filler metal. *Mater. Des.* **2013**, *47*, 717–724. [CrossRef]
2. Hangai, Y.; Kamada, H.; Utsunomiya, T.; Kitahara, S.; Kuwazuru, O.; Yoshikawa, N. Aluminium alloy foam core sandwich panels fabricated from die casting aluminum alloy by friction stir welding route. *J. Mater. Proces Tech.* **2014**, *214*, 1928–1934. [CrossRef]

3. Wei, D.; Songbai, X.; Bo, S.; Jiang, L.; Suiqing, W. Study on microstructure of 6061 aluminum alloy brazed with Al-Si-Zn filler metals bearing Sr and Ti. *Rare Met. Mater. Eng.* **2013**, *42*, 2442–2446. [CrossRef]
4. Dai, W.; Xue, S.; Lou, J.; Lou, Y.; Wang, S. Torch barzing 3003 aluminum alloy with Zn-Al filler metal. *Trans. Nonferrous Met. Soc. China* **2012**, *22*, 30–35. [CrossRef]
5. Huang, Y.; Gong, J.; Lv, S.; Leng, J.; Li, Y. Fluxless soldering with surface abrasion for joining metal foams. *Mater. Sci. Eng. A* **2012**, *552*, 283–287. [CrossRef]
6. Wan, L.; Huang, Y.; Huang, T.; Lv, S.; Feng, J. Novel method of fluxless soldering with self-abrasion for fabricating aluminum foam sandwich. *J. Alloys Compd.* **2015**, *640*, 1–7. [CrossRef]
7. Ubertalli, G.; Ferraris, M.; Bangash, M.K. Joining of AL-6016 to Al-foam using Zn-based joining materials. *Compos. Part A Appl. Sci. Manuf.* **2017**, *96*, 122–128. [CrossRef]
8. Bangash, M.K.; Ubertalli, G.; Di Saverio, D.; Ferraris, M.; Jitai, N. Joining of aluminium alloy sheets to aluminium alloy foam using metal Glasses. *Metals* **2018**, *8*, 614. [CrossRef]
9. Hu, S.P.; Niu, C.N.; Bian, H.; Song, X.G.; Cao, J.; Tang, D.Y. Surface activation assisted brazing of Al-Zn-Mg-Cu alloy: Improvement in microstructure and mechanical properties. *Mater. Lett.* **2018**, *218*, 86–89. [CrossRef]
10. Schällibaum, J.; Burbach, T.; Münch, C.; Weiler, W.; Wahlen, A. Transient liquid phase bonding of AA 6082 aluminium alloy. *Materialwiss. Werkst.* **2015**, *46*, 704–712. [CrossRef]
11. Qingxian, H.; Sawei, Q.; Yuebo, H. Development on preparation technology of aluminum foam sandwich panels. *Rare Met. Mat. Eng.* **2015**, *44*, 0548–0552. [CrossRef]
12. Xiao, B.; Wang, D.; Cheng, F.; Wang, Y. Oxide film on 5052 aluminum alloy: Its structure and removal mechanism by activated CsF-AlF3 flux in brazing. *Appl. Surf. Sci.* **2015**, *337*, 208–215. [CrossRef]
13. Zhu, Z.; Chen, Y.; Luo, A.A.; Liu, L. First conductive atomic force microscopy investigation on the oxide film removal mechanism by chloride fluxes in aluminum brazing. *Scr. Mater.* **2017**, *138*, 12–16. [CrossRef]
14. Boonyongmaneerat, Y.; Schuh, C.A.; Dunand, D.C. Mechanical properties of reticulated aluminum foams with electrodeposited Ni-W coatings. *Scr. Mater.* **2008**, *59*, 336–339. [CrossRef]
15. *ASM Handbook. Properties and Selection: Nonferrous Alloys and Special-Purpose Materials*; ASM International: Russell Township, OH, USA, 1990; Volume 2.
16. Gancarz, T.; Pstrus, J.; Fima, P.; Mosinska, S. Effect of Ag addition to Zn-12Al alloy on kinetics of growth of intermediate phases on Cu substrate. *J. Alloy. Compd.* **2014**, *582*, 313–322. [CrossRef]
17. Contreras, A.; Bedolla, E.; Perez, R. Interfacial phenomena in wettability of TiC by Al-Mg alloys. *Acta Mater.* **2004**, *52*, 985–994. [CrossRef]
18. Contreras, A. Wetting of TiC by Al-Cu alloys and interfacial characterization. *J. Colloid Interface Sci.* **2007**, *311*, 159–170. [CrossRef] [PubMed]
19. Swidersky, H.-W. Aluminium brazing with non-corrosive fluxes state of the art and trends in NOCOLOK® flux technology. In Proceedings of the 6th International Conference on Brazing, High Temperature Brazing and Diffusion Bonding, Aachen, Germany, 8–10 May 2001.

© 2019 by the authors. Licensee MDPI, Basel, Switzerland. This article is an open access article distributed under the terms and conditions of the Creative Commons Attribution (CC BY) license (http://creativecommons.org/licenses/by/4.0/).

Article

Grafting of Gallic Acid onto a Bioactive Ti6Al4V Alloy: A Physico-Chemical Characterization

Martina Cazzola [1,†], Sara Ferraris [1,†], Enrico Prenesti [2], Valentina Casalegno [1] and Silvia Spriano [1,*]

1. Department of Applied Science and Technology, Politecnico di Torino, C.so Duca degli Abruzzi 24, 10129 Torino, Italy; martina.cazzola@polito.it (M.C.); Sara.ferraris@polito.it (S.F.); Valentina.casalegno@polito.it (V.C.)
2. Department of Chemistry, Università degli Studi di Torino, Via Pietro Giuria 5, 10125 Torino, Italy; enrico.prenesti@unito.it
* Correspondence: silvia.spriano@polito.it; Tel.: +39-1-1090-5768
† These authors contribute equally to this paper.

Received: 2 April 2019; Accepted: 25 April 2019; Published: 3 May 2019

Abstract: Despite increasing interest in the use of natural biomolecules for different applications, few attempts of coupling them to inorganic biomaterials are reported in literature. Functionalization of metal implants with natural biomolecules could allow a local action, overcoming the issue of low bioavailability through systemic administration. In the present work, gallic acid was grafted to a pre-treated Ti6Al4V in order to improve its biological response in bone contact applications. The grafting procedure was optimized by choosing the concentration of gallic acid (1 mg/mL) and the solvent of the solution, which was used as a source for functionalization, in order to maximize the amount of the grafted molecule on the titanium substrate. The functionalized surfaces were characterized. The results showed that functionalization with Simulated Body Fluid (SBF) as solvent medium was the most effective in terms of the amount and activity of the grafted biomolecule. A key role of calcium ions in the grafting mechanism is suggested, involving the formation of coordination compounds formed by way of gallic acid carboxylate and Ti–O$^-$ as oxygenated donor groups. Bioactive behavior and surface charge of the pre-treated Ti6Al4V surface were conserved after functionalization. The functionalized surface exposed a greater amount of OH groups and showed higher wettability.

Keywords: titanium; gallic acid; polyphenols; surface functionalization; metal implants

1. Introduction

Polyphenols are organic molecules attracting more and more interest within the scientific community and their application is studied in different fields like medicine, pharmacology, packaging, food conservation and cosmetics. Polyphenols have anti-inflammatory, antioxidant, anticancer and antibacterial effects [1–3]. They can also affect bone health, i.e., stimulating differentiation of healthy osteoblasts [4] and promoting the apoptosis of osteosarcoma cells [5]. These molecules are also able to influence bone density [6,7] and the in vitro deposition of hydroxyapatite [8,9].

Surface functionalization and coating with natural biomolecules is a material science emerging field: Some attempts of coupling polyphenols to biomaterials are reported in literature till now and they are briefly summarized below.

As far as nanoparticles functionalization is concerned, benzoic, caffeic, coumaric, ferulic and syringic acids from plant extracts were linked to magnetic nanoparticles using polyethylene glycol (PEG) as linker molecule [10]; quercetin was covalently linked to rare earth nanoparticles by means of silane [11]; resveratrol was coupled with gold [12], polycaprolactone [13] and polycaprolactone

(PCL)/collagen nanoparticles [4] and gallic acid were encapsulated in zein electrospun nanoparticles [14]. Concerning nanostructured substrates, apple polyphenols were encapsulated in a β-cyclodextrin nanostructured sponge [15], rutine (present in plants such as those of the genus *Citrus*) was adsorbed on mesoporous silica mesostructured MCM-41 type [16] and polyphenols from green tea leaves were coupled with carbon nanotubes [17].

Considering polymers functionalization, the following biomolecules were grafted to fibers, films or microspheres of chitosan: Flavonoids (by means of the tyrosine kinase enzyme) [18], quercetin, tannic acid (by means of laccase enzyme) [19], caffeic acid [20], thyme polyphenols [21], Zataria multiflora Boiss essential oil, extract of grape seeds [22], raspberry leaf, hawthorn, ivy, yarrow, nettle and olive leaves [23]. Curcumin was physically encapsulated in polyurethane membranes [24]. Epigallocatechin gallate (EGCG) was used for functionalization of collagen fibers [25] while extract of coriander was physically encapsulated in a bone graft material (biphasic calcium phosphate and casein chitosan) [26].

Few researches considered inorganic materials as substrates for functionalization: Extract of Gusuibu (*Drynaria roosii*) was covalently linked to calcium hydrogen phosphate [27]; natural polyphenols and pyrogallol were used as coating on stainless steel [28]. Silica-based bioactive glasses and glass ceramics were successfully functionalized with gallic acid and natural extracts from red grape skins and green tea leaves by the authors, as reported in previous papers [29,30].

New protocols for functionalization have to be developed if we change the substrate moving towards titanium alloys. Surface modifications of titanium and its alloys, in order to confer bioactive and antibacterial properties, are widely studied [31–33], but papers related to functionalization with natural biomolecules are very few in number. The following biomolecules were coupled to titanium or titanium dioxide substrates: Natural polyphenols and pyrogallol [28], quercetin [34], taxifolin [35], rhamnogalacturonan-Is (RG-Is) isolated from potatoes and apples [36], lignin [37] and gallic acid esters (namely, octyl-, decyl-, lauryl- and cetyl-gallate) [38].

In this work, a surface functionalization protocol has been optimized (selection of solvent medium and pH, effects of Ca^{2+} ions in solutions, concentration of the source solution) in order to graft a significant amount of gallic acid (GA) onto Ti6Al4V alloy surface, previously made bioactive through a specific chemical treatment. Gallic acid (3,4,5-trihydroxybenzoic acid) is used here as a model biomolecule for polyphenols. It shows antioxidant, neuroprotective and antitumor abilities [39,40]. A preliminary characterization of the functionalized surfaces was performed by means of the Folin–Ciocalteu photometric test and XPS (X-ray Photoelectron Spectroscopy) analysis, to verify the presence and activity of the grafted GA. The in vitro bioactivity of the samples was investigated by soaking in Simulated Body Fluid (SBF). Surface wettability and surface charge were evaluated by means of contact angle and zeta potential measurements, respectively.

2. Materials and Methods

2.1. Surface Activation

Ti6Al4V disks (2 mm in thickness, 10 mm diameter, ASTM B348 [41], Gr5, Titanium Consulting and Trading) were polished with SiC paper (up to 4000 grid) and washed in an ultrasonic bath (5 min in acetone and twice 10 min in ultrapure water).

In order to induce bioactive behavior and high density of hydroxyl groups on the surface, suitable for the further surface functionalization, the titanium substrate underwent a specific patented chemical treatment [42,43]. It includes a first acid etching in hydrofluoric acid and a subsequent controlled oxidation. After this treatment, the samples were exposed for 1 h to UV in order to make the surface more reactive [44]. From now on, the samples treated as above described will be named as "chemically-treated—CT".

2.2. Surface Functionalization

Gallic acid (3,4,5-Trihydroxybenzoic acid, GA, 97.5–102.5% titration, G7384, Sigma-Aldrich, St. Louis, MO, USA) was employed for functionalization as model molecule for polyphenols. This molecule was selected because of its simple structure and extensive literature available on it.

For surface functionalization, source solutions with two different concentrations of the biomolecule (1 and 2 mg/mL) in 3 media (namely, ultrapure water—W, Phosphate Buffered Saline—PBS and Simulated Body Fluid—SBF) were prepared. PBS (PBS, Sigma-Aldrich, P4417) was prepared by dissolving 1 tab (2 g) in 200 mL of ultrapure water, while SBF was prepared according to the protocol described by Kokubo [45].

Each sample, after the above described chemical treatment and UV irradiation (CT), was put in a holder with 5 mL of the selected GA solution in the dark for 8 h at 37 °C. After the incubation, the samples were washed two times with a quick immersion in ultrapure water, dried at room temperature and preserved in the dark. The nomenclature of the source solutions and of the functionalized samples is reported in Table 1.

Table 1. Acronyms and a brief description of the source solutions and of the functionalized samples.

Sample Acronym	Sample Description
WATER + GA_1	Solution 1 mg/mL of Gallic acid in ultrapure water
WATER + GA_2	Solution 2 mg/mL of Gallic acid in ultrapure water
SBF + GA_1	Solution 1 mg/mL of Gallic acid in SBF
SBF + GA_2	Solution 2 mg/mL of Gallic acid in SBF
PBS + GA_1	Solution 1 mg/mL of Gallic acid in PBS
PBS + GA_2	Solution 2 mg/mL of Gallic acid in PBS
CT	Ti6Al4V sample chemical treated and UV irradiated
CT_GA_W_1	CT sample functionalized with solution 1 mg/mL of Gallic acid in ultrapure water
CT_GA_W_2	CT sample functionalized with solution 2 mg/mL of Gallic acid in ultrapure water
CT_GA_SBF_1	CT sample functionalized with solution 1 mg/mL of Gallic acid in SBF
CT_GA_SBF_2	CT sample functionalized with solution 2 mg/mL of Gallic acid in SBF
CT_GA_PBS_1	CT sample functionalized with solution 1 mg/mL of Gallic acid in PBS
CT_GA_PBS_2	CT sample functionalized with solution 2 mg/mL of Gallic acid in PBS

2.3. Detection of the Grafted Biomolecule

The presence and activity of GA on the surface of the functionalized samples were investigated by means of the Folin–Ciocalteu test (F&C) and by means of X-ray Photoelectron Spectroscopy (XPS).

The F&C test measures the redox reactivity of phenolic substances and give a quantification of their amount using a calibration curve obtained with GA as standard [29]. The test was performed on the source solution (before and after functionalization) and on the functionalized samples as reported in previous works [29,30,46]. This analysis was performed in triplicate and statistical analysis was obtained through ANOVA one-way test ($p < 0.05$).

As far as XPS analyses (XPS, PHI 5000 VERSAPROBE, Physical Electronics, Chigasaki, Japan) are concerned, both the survey spectra and high-resolution analyses of carbon and oxygen regions were performed to determine the chemical composition of the surface, the elemental chemical state of and to identify the functional groups exposed on the surface. The results were used to define the best functionalization medium and the best concentration of the source solution for optimization of the functionalization protocol.

2.4. Contact Angle Measurements

In order to evaluate surface wettability, contact angle measurements were performed by the sessile drop method before and after functionalization with ultrapure water. A heating microscope (Misura®, Modena, Italy, Expert System Solutions) at room temperature, was used for the acquisition of images and the contact angle value was obtained by images elaboration with Image J (version 1.47) software.

2.5. In Vitro Apatite-Forming Ability Tests (Bioactive Behavior)

The apatite-forming ability of the samples before and after functionalization was investigated by soaking the samples in SBF. Two specimens for each type of sample were soaked in the dark in 25 mL of SBF, prepared according to the protocol developed by Kokubo and Takadama [47], at 37 °C up to 28 days.

2.6. Field Emission Scanning Electron Microscopy (FESEM) Observations and Energy Dispersive Spectroscopy (EDS) Analyses

After soaking in SBF, eventual deposition of hydroxyapatite on the surfaces was observed by means of Field Emission Scanning Electron Microscopy equipped with Energy Dispersive Spectroscopy (FESEM-EDS SUPRATM 40, Zeiss and Merlin Gemini Zeiss, Cohen, Germany). Samples have been observed without metallization.

2.7. Electro-Kinetic Measurements

The zeta potential of the samples in function of pH was analyzed by means of electro-kinetic measurements (SurPASS, Anton Paar, Graz, Austria) using 0.001 M KCl as electrolyte. Basic and acid titrations are two separate measurements performed varying the pH by the addition of 0.05 M HCl or 0.05 M NaOH, respectively, through the automatic titration unit of the instrument.

3. Results and Discussion

3.1. pH Measurements

The pH values of the solutions before (source solutions) and after (uptake solutions) the procedure of functionalization were measured and the color was observed (Table 2).

Table 2. pH values of the source and uptake solutions.

Solution	pH	Color of the Solution
WATER + GA_1 (Source)	3.35 ± 0.04	Colorless
WATER + GA_2 (Source)	3.17 ± 0.04	Colorless
SBF + GA_1 (Source)	7.45 ± 0.10	Blue
SBF + GA_2 (Source)	7.10 ±0.11	Blue
PBS + GA_1 (Source)	6.17 ±0.09	Colorless
PBS + GA_2 (Source)	4.60 ±0.12	Colorless
WATER + GA_1 (Uptake)	3.36 ± 0.01	Colorless
WATER + GA_2 (Uptake)	3.15 ± 0.04	Light yellow
SBF + GA_1 (Uptake)	7.22 ± 0.04	Blue
SBF + GA_2 (Uptake)	6.71 ± 0.08	Blue
PBS + GA_1 (Uptake)	6.05 ± 0.07	Light yellow
PBS + GA_2 (Uptake)	4.49 ± 0.04	Colorless

The pH values and color are significant parameters for the solutions containing polyphenols, because they give a first indication of the chemical state of the biomolecules. The electronic absorption spectra of polyphenols are strongly dependent on the medium used for their dissolution, the presence of electro-donating of electron-withdrawing substituents such as metal ions and formation of a pH dependent resonance form [48].

In order to allow the binding of a significant amount of biomolecules to the surface, the protocol of functionalization used in this research was developed from previous works of the authors [29,30] and from literature data [28]. The previous works of the authors concern surface functionalization of bioactive glasses and glass ceramics. A protocol of functionalization was elaborated over time in order to optimize grafting of the biomolecules on the glasses avoiding degradation of the biomolecules themselves. In the case of bioactive glasses, the main issue is the regulation of pH avoiding a too strong alkalization induced by massive ion release from bioactive glasses in water-based media. This work is focused on titanium surfaces and a different protocol must be developed.

In this case, ion release from the substrate into water-based media is almost negligible and consequently pH of the functionalization solution during sample soaking remains close to the initial one. In this case, in order to maximize the reactivity of polyphenols and their ability to be bond to the surface, both pH and ionic composition of the solution should be optimized in order to favor surface grafting. This specific point is the topic of the present research work.

The source water solutions of GA were slightly acidic (pH is around 3) and they did not change in a significant way after incubation of the samples. These solutions were colorless, due to the acidic pH and the absence of metal cations (necessary for the formation of colored complexes with GA) [48,49]. The uptake solution WATER + GA_2 evidenced a slight change of color, turning from colorless to light yellow. This phenomenon can be associated with the release of few Ti^{4+} ions in the solution and the consequent formation of gallic acid titanium complexes which can be yellow colored [50]. Within this range of pH, deprotonation of GA is not enhanced significantly. PBS and SBF were utilized as alternative media for the source solutions of GA in order to perform functionalization at pH values higher than 4: Pure PBS and SBF media have both a pH value of 7.40 [51,52]. At this pH, deprotonation of the carboxylic group of the biomolecule is induced and grafting to the surface can be improved [53]. However, pH of the source solution should not be higher than 10 to avoid immediate and irreversible oxidative degradation (shift from phenol to quinone groups) of GA. It is also important to set and control the time of incubation, because a long permanence at pH values exceeding 7.40 leads to a slow degradation of the molecule [48].

For the PBS solutions, the pH values of the source solutions PBS + GA_1 and PBS + GA_2 were 6.17 and 4.60 respectively and they remained almost unchanged in the uptake solutions. The source PBS + GA solutions were colorless, which can be attributed both to pH which was lower than 7.40 upon GA additions, and to the absence of metal cations [48,49]. For the uptake solution with 1 mg/mL of GA (Uptake PBS + GA_1), a light-yellow color appeared after functionalization: This was not completely unexpected, because a small oxidative degradation of gallic acid can be caused by a number of environmental factors. It must be also underlined that PBS did not have an effective buffer action on pH and a clear trend in behavior of PBS solution was not observed.

In the SBF solutions, the deprotonation of carboxylic acid was also improved by the presence of calcium cations that were not present in PBS solution. The source solution SBF + GA_2 showed a slight decrease of pH down to 7.10 with respect to pure SBF because of the addition of GA. The source and uptake solutions with a GA concentration of 2 mg/mL show a higher decrease of pH: This was due to the tendency of GA, which was present in a higher amount, to dissociate, producing H^+ ions.

All the solutions of GA in SBF were blue because of the deprotonation of the carboxylic group and the tendency of the polyphenols to form complexes with metal cations (e.g., Ca^{2+}). The investigation of the role of calcium in functionalization of surfaces with polyphenols was as first preliminary performed by using a simple water medium containing only $CaCl_2$ (with Ca^{2+} ions in the same concentration of SBF), a solution of buffer (Tris(hydroxymethyl)aminomethane (TRIS)/HCl) and a solution containing both. It has been observed that the presence of Ca^{2+} ions is crucial for GA binding on CT surfaces. In fact, the use of a buffer (TRIS/HCl) without Ca^{2+} is not sufficient for an effective functionalization (unpublished results). The preliminary test performed evidenced a blue coloration of the functionalization solution prepared with $CaCl_2$ and TRIS/HCl buffer which suggest the formation of GA-Ca^{2+} complexes in the solution. The first functionalization trial of CT surfaces with these solutions suggests a crucial role of GA-Ca^{2+} complexes for an effective grafting of GA to titanium surfaces and it was investigated in details in this work.

The time of incubation is a constant throughout this work (8 h) and it was selected as a compromise among different issues: Maximization of the amount of the grafted biomolecule, reasonable process time, avoiding corrosion of the substrate in the case of source solutions prepared with water as medium (pH around 3) and avoiding degradation of the biomolecule at physiological pH (close to 7.4) in the case of source solutions prepared with PBS and SBF as media.

3.2. Biomolecule Detection

Figure 1a shows the concentration of GA in the source solutions quantified by way of the F&C test. The values obtained were comparable with the nominal concentrations, confirming that GA was stable in the chosen solvents, at least for the time necessary for the selected protocol. Only the solution PBS + GA_2 shows a concentration lower than the nominal one and it could indicate an initial degradation of the biomolecule not evidenced by a color change. The analyses on the uptake solutions did not highlight a significant lowering of the concentration of GA compared to that of the source solutions: The amount of the grafted GA was too low to be detected by this way. The photometric analyses performed on the functionalized samples showed the presence of GA on all the surfaces. As first, the tendency of the polyphenols to bind to the surfaces exposed to aqueous solutions reported in literature was confirmed [28,54].

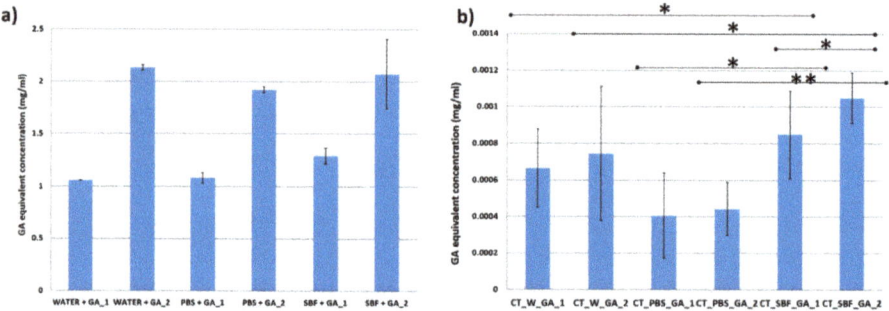

Figure 1. Results of the Folin–Ciocalteu test for quantification of gallic acid in (**a**) source solutions and on (**b**) functionalized samples (* $p < 0.05$, ** $p < 0.5$).

Figure 1b highlights a statistically significant greater amount of GA grafted on the samples functionalized in SBF + GA with respect to the samples WATER + GA and the samples PBS + GA ($p < 0.05$) comparing the samples treated with the same GA concentration. The difference of polyphenol concentration on the surfaces treated with the two different concentrations of GA in the same solvent has a low statistical significance ($p > 0.05$). This observation suggests that, in the reported functionalization conditions, the surface reactive sites, available for GA grafting, are almost completely saturated upon contact with 1 mg/mL solutions.

The atomic percentages of the elements detected on the surface of the Ti6Al4V samples by means of XPS survey analyses, after functionalization from different source solutions, are reported in Table 3.

Table 3. Atomic percentages of the elements detected on the samples by XPS survey analyses. (Uncertainty of measurements 0.5 at.%–1 at.%).

Elements [at.%]	Samples						
	CT	CT_GA_W_1	CT_GA_W_2	CT_GA_PBS_1	CT_GA_PBS_2	CT_GA_SBF_1	CT_GA_SBF_2
O	57.0	52.6	49.6	59.0	55.5	44.7	47.8
C	19.0	29.9	31.8	19.1	20.2	45.2	39.1
Ti	18.2	15.9	14.9	15.6	15.6	6.6	8.7
Ca	–	–	–	0.8	–	3.5	3.2
Others	5.8	1.6	3.8	–	5.4	–	8.7

A certain amount of carbon can be detected on the CT samples due to unavoidable contaminations on the titanium surfaces [55]. A moderate increase of the carbon content can be observed on the samples functionalized in water solutions (CT_GA_W_1 and CT_GA_W_2) and a higher one on the samples functionalized in SBF solutions (CT_GA_SBF_1 and CT_GA_SBF_2). On the other hand, a negligible variation of the surface carbon content can be detected on the samples functionalized in PBS (CT_GA_PBS_1 and CT_GA_PBS_2). The increase of the surface carbon can be correlated to grafting of GA and the results are in agreement with the results of the Folin–Ciocalteu test.

A not negligible amount of calcium ion was detected on the samples functionalized in SBF (CT_GA_SBF_1, CT_GA_SBF_2) while negligible amount of calcium ion was observed on one sample functionalized in PBS (CT_GA_PBS_1): In this case, due to the very low amount and absence in PBS solution (at least theoretically), it can be considered a contamination. The presence of calcium on the samples incubated in the source solutions with SBF as medium, can be related to the GA ability to bind this element [8,30] within a complexation reaction [49,56]. It is reported [48,56] that polyphenols are prone to give complexation reactions in solutions containing metal ions; the structure of the coordination compounds can be different depending on the pH of the solution. At pH values higher than 6 can be hypothesized that a heterogeneous ternary coordination compound is formed with the metal ions in a central position. The complex compound formation can involve both the carboxylate ion of GA and Ti–O$^-$ as donor groups. The formation of the coordination compound influences the chemical reactivity of polyphenols, for instance they are able to polymerize and to form coatings [56,57]: Particularly, in the present case it is shown that they were much more prone to be grafted on a surface. The possible formation of the complex and its binding to the surface is reported in Figure 2.

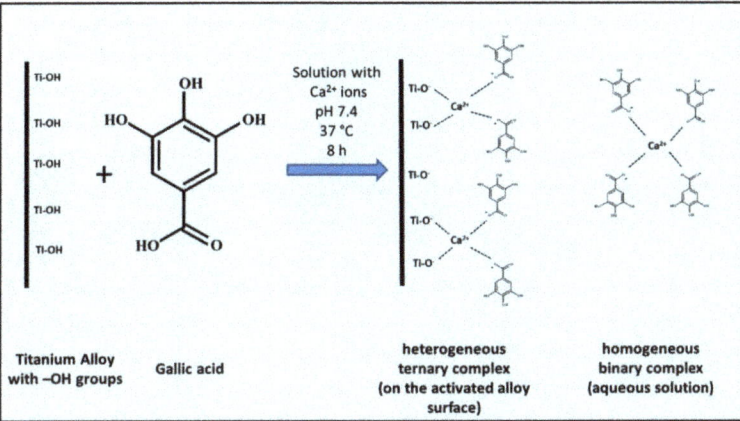

Figure 2. Scheme of the possible binding of gallic acid with the titanium alloy surface and Ca (II). Only the planar coordination is reported.

At pH 7.40, the GA (in the gallate chemical form, with the carboxylic group deprotonated) was bound to the alloy surface (in the Ti–O$^-$ chemical form) having Ca^{2+} ion as linking species. A heterogeneous ternary complex was then probably formed, allowing the grafting of the gallate ion to the activated alloy surface). In the uptake solution, the excesses GA and free calcium ions could form homogeneous complexes (Ca(II)-gallate with different coordination numbers), in Figure 2 one of the possible planar coordination is reported, the eventual equatorial coordination and 3D structure is omitted. The presence of calcium ions could also allow a multistep assembly of polyphenols coordination obtaining a thicker layer of polyphenols grafted to the surfaces in order to meet the threshold concentration needed to induce beneficial effects on cells.

The ability of polyphenols to attract and bind calcium ions could have an influence also in the induced precipitation of hydroxyapatite during soaking in SBF (see Section 3.4) [56] as already observed by the authors in the case of surface functionalization of bioactive glasses with GA [30]. The absence of calcium in the PBS medium can explain the significantly lower amount of GA grafted on the surfaces exposed to source solutions prepared with this medium: The presence of a metal ion forming complex with polyphenols is crucial for their grafting.

The observation of the surface chemical composition gave some information, but it was not sufficient in order to determine the presence of the grafted biomolecule. In order to detect the specific functional groups of GA, a detailed analysis of the carbon and oxygen regions was performed and the results are reported in Figures 3 and 4.

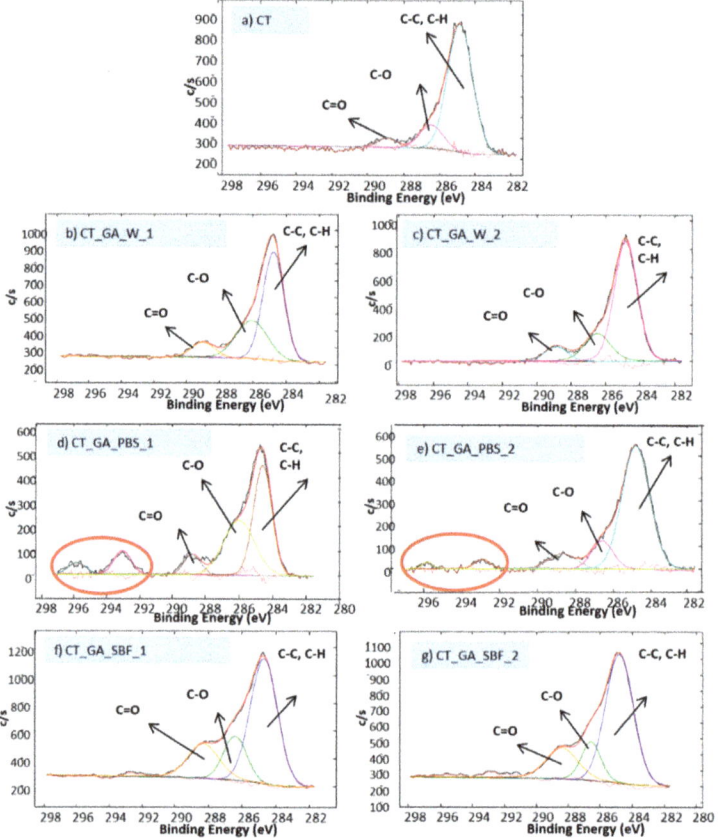

Figure 3. XPS high-resolution spectra of carbon region. (**a**) CT; (**b**) CT_GA_W_1; (**c**) CT_GA_W_2; (**d**) CT_GA_PBS_1; (**e**) CT_GA_PBS_2; (**f**) CT_GA_SBF_1; (**g**) CT_GA_SBF_2.

Looking at the high-resolution spectra of the carbon region of the CT samples (Figure 3a), the main signal at 284.81 eV can be observed together with two small contributions at 286.26 and 289.17 eV. The signal at 284.81 eV can be attributed to C–C and C–H bonds within hydrocarbon contaminants, always present on the titanium surfaces which are very reactive towards carbon and it was already detected by the authors [31]. The other two contributions respectively fall in the regions of C–O and C=O bonds [57,58]: they have low intensities on this sample and are associated to surface contaminations [31]. The signal at 284.81 eV is unchanged on the functionalized samples (Figure 3b–g). On the other hand, a significant increase in the contributions at 286.26 and 289.17 eV is evident for CT_GA_SBF_1 and CT_GA_SBF_2. As far as the signal at 284.81 eV is concerned, C–C and C–H bonds are typical of the above-cited hydrocarbon contamination but are also present within the GA molecule. The presence of C–O and C=O can be associated with the presence of GA and to its tendency (phenol groups) to oxidize into quinone, as previously observed by the authors [29,30].

A further couple of signals at about 293 and 295.8 eV can be observed on the samples functionalized in PBS (CT_GA_PBS_1 and CT_GA_PBS_2, Figure 3d,e). They can be associated to shake up satellite peaks due to aromatic rings [59,60] and suggest a different disposition of GA molecules grafted on the titanium surfaces when PBS is used as medium for the source solutions. The different orientation of GA grafted on the surface after functionalization in PBS can be explained considering (i) the different pH value (in comparison with that of the source solutions prepared in water); and ii) the chemical composition, with the absence of the Ca^{2+} ion with its coordinating ability towards oxygen donor groups, (in comparison with that of the source solutions prepared using SBF as solvent). The different orientation of the grafted biomolecule can be related to the lower redox activity registered by the Folin & Ciocalteu test on these surfaces.

The high-resolution spectra of the oxygen region are reported in Figure 4.

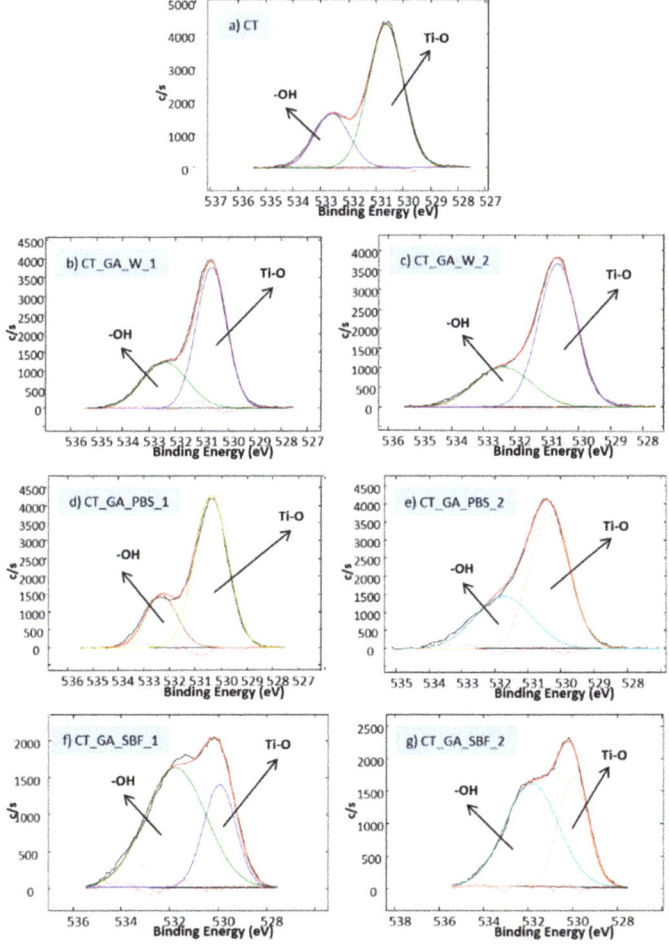

Figure 4. XPS high-resolution spectra of the oxygen region. (**a**) CT; (**b**) CT_GA_W_1; (**c**) CT_GA_W_2; (**d**) CT_GA_PBS_1; (**e**) CT_GA_PBS_2; (**f**) CT_GA_SBF_1; (**g**) CT_GA_SBF_2.

Two signals around 530 and 532 eV can be observed on all the samples. The first one can be attributed to oxygen within the Ti–O bonds in the titanium oxide layer, while the second one can be assigned to the OH groups [43]. OH groups are more abundant on the surface chemically treated (CT) samples than on pristine TI6Al4V alloy, as previously evidenced by the authors [43,61,62]. A further notable increase in the OH signal can be observed after GA grafting, mainly on the samples functionalized by using the source solutions with SBF as medium, and it can be explained considering that GA is rich in hydroxyl groups and they are still exposed after grafting. In the case of the samples functionalized in a source solution with PBS as medium, the lower amount of OH groups can be related both to a reduced amount of grafted biomolecule and to a different orientation on the surface. The growth of the OH peak after gallic acid functionalization was already observed by the authors on the functionalized bioactive glasses [29,30]. Since functionalization starting from source solutions with SBF as medium gave the best results in terms of amount and activity of grafted GA and that no highly significant ($p < 0.05$) differences were detected between the two tested concentrations, further analyses were performed only on the CT_GA_SBF_1 samples and on the CT samples as reference.

3.3. Contact Angle Measurements

The bar graph in Figure 5 shows the contact angle measured of CT and CT_GA_SBF_1.

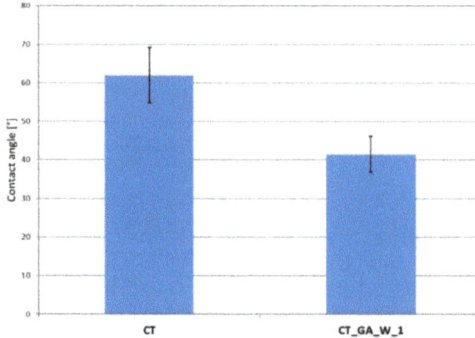

Figure 5. Contact angle measurement before and after the functionalization.

The contact angle was 62° ± 7° for CT and 41° ± 5° for CT_GA_SBF_1. The lowering of the contact angle value can be related to the presence of the grafted molecules of GA exposing its hydroxyl groups, as confirmed by XPS data.

3.4. In Vitro Apatite-Forming Ability Tests

The pH measurements during the soaking period highlighted a small decrease of the pH of the SBF solution CT_GA_SBF_1 samples during the first 3 days of soaking, probably correlated to a partial release of the grafted molecules from the surface, but also to the tendency of a portion of OH groups on the surfaces to dissociate into H^+ ions. During the further refreshes, no significant changes of the pH were observed.

The FESEM micrographs of the samples after 28 days in SBF are reported in Figure 6.

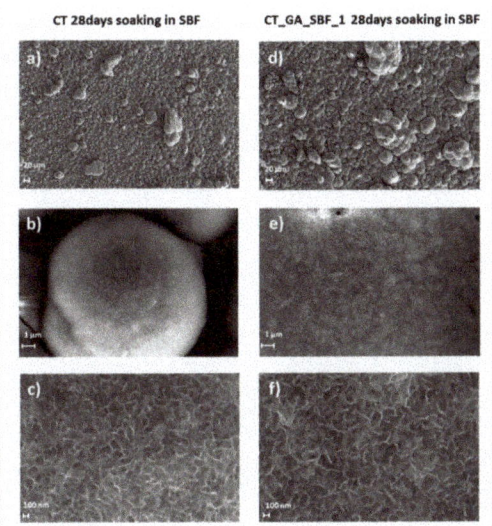

Figure 6. FESEM images of samples after 28 days SBF soaking at different magnifications. (**a**). (500×), (**b**) (20k×) and (**c**) (100k×) for CT sample, (**d**) (500×), (**e**) (20k×) and (**f**) (100k×) for CT_GA_SBF_1 one.

The surfaces of the functionalized and reference samples are covered by hydroxyapatite with the usual cauliflower-like morphology. The CT samples have good in vitro bioactivity as reported in a previous work [50] and on the functionalized samples the amount of grafted GA molecules did not significantly modify the bioactive behavior. In Table 4 the results of the EDS analyses performed on different areas (106400 μm^2) of the samples are reported.

Table 4. Atomic percentages of the elements detected on the samples by EDS survey.

Elements (at.%)	Samples	
	CT	CT_GA_SBF_1
C	7.54 ± 3.06	7.59 ± 2.96
O	55.10 ± 10.85	55.09 ± 10.86
Na	0.48 ± 0.03	0.45 ± 0.03
Mg	0.73 ± 0.07	0.75 ± 0.07
P	13.32 ± 3.54	12.71 ± 3.37
Ca	22.18 ± 9.68	22.18 ± 9.67
Ti	0.78 ± 0.42	1.25 ± 0.67
Ca/P ratio	1.63 ± 0.3	1.7 ± 0.3

Calcium and phosphorus were present on the surfaces of both the samples and their ratio (Ca/P), 1.63 for CT samples and 1.70 for CT_GA_SBF_1 samples, were comparable with the stoichiometric one of hydroxyapatite (1.67) [63].

The kinetic of the hydroxyapatite formation as a function of the surface functionalization will be investigated in a further work to clarify if the ability of GA to react with calcium can affect the kinetic of bioactive behavior, besides the mechanism of grafting of the biomolecule, as discussed above. From these results it is possible to say that the functionalization did not decrease the bioactivity of the samples and further cellular tests are needed in order to understand the effects on cells.

3.5. Zeta Potential Measurements

The zeta potential titration curve in function of the pH of the solution is reported in Figure 7 in the case of polished surfaces (reported in Figure 7 as Ti6Al4V), CT samples, CT samples soaked 28 days in SBF, CT_GA_SBF_1 samples and CT samples soaked 28 days in SBF.

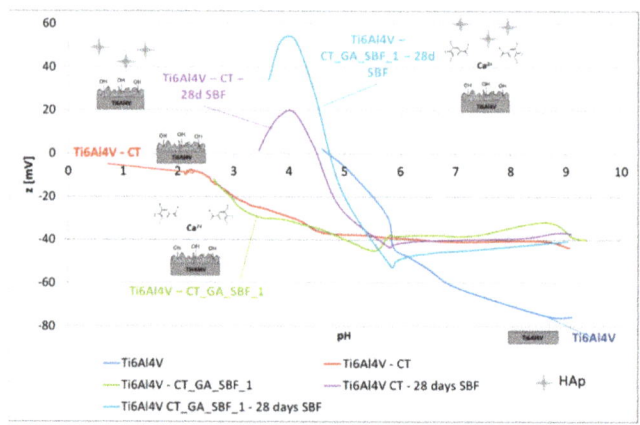

Figure 7. Zeta potential recorded as function of pH.

Some curves show a discontinuity in the range of pH between 5.5–6 and it was an artefact due to the apposition of the acid and basic sides of the titration curve which were obtained in different times, indeed if the discontinuity is lower than 10 mV, the surface alteration is considered negligible, otherwise different samples should be used for the acid and basic titrations.

An isoelectric point of 4.7 has been obtained on the polished surface, in accordance to the value reported in literature for titanium [64]. A shift of the isoelectric point to more acidic values (close to 2, extrapolating the curves in Figure 7) can be observed in the case of the CT and CT_GA_SBF_1 samples with no significant differences. The shift to a more acidic value of the isoelectric point after surface oxidation of titanium is not obvious. For instance, Kulkarni et al. [64] reported a basic shift of the isoelectric point in the case of anodized titanium (with stoichiometric titanium oxide nanotubes) compared to a titanium foil (measurement carried on the amorphous native oxide). The acidic shift of the isoelectric point here observed on the CT and CT_GA_SBF_1 samples compared to the polished surface can be attributed to the enrichment in acidic hydroxyl groups due to the presence of the chemically modified oxide layer (CT) and to the presence of GA (CT_GA_SBF_1). According to their acidic isoelectric points (2–4.7) at physiological pH, all the considered surfaces were negatively charged, however, some differences can be pointed out between the samples. At first, a notable reduction in the absolute value of the surface charge can be observed for the CT and CT_GA_SBF_1 samples (≈ -40 mV), compared to the polished one (≈ -75 mV). Moreover, a plateau for the surface charge can be observed for the CT and CT_GA_SBF_1 samples and not for the polished one. This plateau can be ascribed to the presence of homogeneous chemical groups exposed on the surface, in this case acidic OH groups, as evidenced in [65], which were completely dissociated at the onset of the plateau. This result was in accordance with the acidic shift of the isoelectric points and it was a further confirmation of the presence of OH groups on both the CT and CT_GA_SBF_1 surfaces.

The absence of a shift of the isoelectric point (IEP) in CT_GA_SBF_1 with respect to CT was compatible with functionalization with a mono and not continuous layer, as well as to the grafting of a molecule (GA) which have an acidic isoelectric point (not far from the one of CT samples) as reported in literature for various phenolic acids [66].

The curve of the CT_GA_SBF_1 sample showed some distinctive features highlighting that after functionalization there is a change of the surface. There was a small plateau around pH 3–4 that could come from partial ionization of carboxyl groups of the GA bonded to the surface [67]. There was a trend towards positive charge after pH 7.20 with a peak around pH 8.50 that could be connected to the degradation of gallic acid at high alkaline pH or to the desorption of phosphate groups adsorbed during the functionalization in the solution with SBF as medium.

In the case of the curves of the CT and CT_GA_SBF_1 samples soaked for 28 days in SBF for the apatite-forming ability tests, the IEP was moving to pH 5 that is the IEP of hydroxyapatite [68], suggesting the presence of a continuous layer of hydroxyapatite as seen by the FESEM analysis.

Some small differences between the two curves can suggest, as reported in literature, an action of the GA on the deposition of hydroxyapatite [8,30]. In the case of the CT sample (without functionalization), the IEP was at a lower pH with respect to the functionalized sample and the plateau in the alkaline zone was shifted towards higher surface charge; the decreasing trend of the curve at pH lower than the IEP can be related to the decomposition of the hydroxyapatite layer in acidic solutions and it occurs at a lower surface charge in the case of the CT sample, similarly the peak after pH 8 can be related to the decomposition of hydroxyapatite in the alkaline range and it occurs only in the case of the CT sample (without functionalization). All these differences could indicate the precipitation of a more stable hydroxyapatite when GA is grafted on the surface: This hypothesis needs further investigation to be proved.

4. Conclusions

GA was grafted to the surface of a chemically treated bioactive Ti6Al4V alloy conserving its redox activity. Surface functionalization was performed in different media (namely, water, PBS and SBF) and at different GA concentrations (1–2 mg/mL). Functionalization by using source solutions with SBF as medium proved to be the most effective in terms of molecular amount and activity; the source solution and the uptake did not show evidence of degradation; no significant difference between the use of source solutions of different concentrations was registered (in the explored range). A key role of the calcium ions in the grafting mechanism was suggested, involving coordination compounds using both carboxylate ions of GA and Ti–O$^-$ as donor groups. Bioactive behavior conferred to the titanium surface by the chemical pre-treatment was conserved after functionalization: Zeta potential measurements suggest a different kinetic of apatite precipitation, but this issue is still subjected to confirmation. The functionalized surface exposed a greater amount of OH groups and showed a slightly higher wettability. These preliminary results show that GA can be effectively grafted to a bioactive Ti6Al4V alloy and are promising for the development of smart metallic biomaterials able to couple the good bulk properties of the titanium alloys with both bioactive surface behavior and the biological benefits and actions of natural biomolecules, locally loaded on the surface of bone implants. The present research can be considered a preliminary study intended for the understanding functionalization mechanisms on bioactive titanium surfaces, by means of physical and chemical characterizations, using gallic acid as a model molecule for polyphenols. Considering the promising results of the present study, surface functionalization of chemically treated titanium surfaces with natural polyphenols (following the developed protocol here) and further biological characterizations should be encouraged as future developments of the work. The possibility to combine the inorganic bioactive behavior of chemically treated titanium substrates with the biological properties of natural polyphenols (antioxidant, anticancer, antibacterial, anti-inflammatory and bone stimulating) is a challenging opportunity for the development of functional biomaterials for bone contact applications in critical situations, by means of an eco-friendly and sustainable approach.

Author Contributions: Conceptualization, all the authors; Methodology M.C., S.F. and S.S.; Investigation, M.C. and S.F.; Resources, S.S.; Data Curation, all the authors; Writing—Original Draft Preparation, M.C. and S.F.; Writing—Review and Editing, E.P., V.C. and S.S.; Supervision, S.S; Software, M.C.; Validation, M.C., S.F. and S.S.; Visualization, S.F.; Project Administration, S.S.

Funding: This research received no external funding.

Conflicts of Interest: The authors declare that they have no conflict of interest.

References

1. Pandey, K.B.; Rizvi, S.I. Plant polyphenols as dietary antioxidants in human health and disease. *Oxid. Med. Cell. Longev.* **2009**, *2*, 270–278. [CrossRef] [PubMed]
2. El Gharras, H. Polyphenols: food sources, properties and applications—A review. *Int. J. Food Sci. Technol.* **2009**, *44*, 2512–2518. [CrossRef]
3. Quideau, S.; Deffieux, D.; Douat-Casassus, C.; Pouységu, L. Plant polyphenols: Chemical properties, biological activities, and synthesis. *Angew. Chem. Int. Ed.* **2011**, *50*, 586–621. [CrossRef] [PubMed]
4. Dosier, C.R.; Erdman, C.P.; Park, J.H.; Schwartz, Z.; Boyan, B.D.; Guldberg, R.E. Resveratrol effect on osteogenic differentiation of rat and human adipose derived stem cells in a 3-D culture environment. *J. Mech. Behav. Biomed. Mater.* **2012**, *11*, 112–122. [CrossRef]
5. Li, Y.; Bäckesjö, C.M.; Haldosén, L.A.; Lindgren, U. Resveratrol inhibits proliferation and promotes apoptosis of osteosarcoma cells. *Eur. J. Pharmacol.* **2009**, *609*, 13–18. [CrossRef]
6. Đudarić, L.; Fužinac-Smojver, A.; Muhvić, D.; Giacometti, J. The role of polyphenols on bone metabolism in osteoporosis. *Food Res. Int.* **2015**, *77*, 290–298. [CrossRef]
7. Ornstrup, M.J.; Harsløf, T.; Kjær, T.N.; Langdahl, B.L.; Pedersen, S.B. Resveratrol increases bone mineral density and bone alkaline phosphatase in obese men: A randomized placebo-controlled trial. *J. Clin. Endocrinol. Metab.* **2014**, *99*, 4720–4729. [CrossRef] [PubMed]
8. Tang, B.; Yuan, H.; Cheng, L.; Zhou, X.; Huang, X.; Li, J. Effects of gallic acid on the morphology and growth of hydroxyapatite crystals. *Arch. Oral Biol.* **2015**, *60*, 167–173. [CrossRef] [PubMed]
9. Zhou, R.; Si, S.; Zhang, Q. Water-dispersible hydroxyapatite nanoparticles synthesized in aqueous solution containing grape seed extract. *Appl. Surf. Sci.* **2012**, *258*, 3578–3583. [CrossRef]
10. Saikia, J.P.; Konwarh, R.; Konwar, B.K.; Karak, N. Isolation and immobilization of Aroid polyphenol on magnetic nanoparticles: Enhancement of potency on surface immobilization. *Colloids Surf. B* **2013**, *102*, 450–456. [CrossRef] [PubMed]
11. Wang, K.; Wu, Y.; Li, H.; Li, M.; Zhang, D.; Huixia, F.; Feng, H.; Fun, H. Dual-functionalization based on combination of quercetin compound and rare earth nanoparticle. *J. Rare. Earths* **2013**, *31*, 709–714. [CrossRef]
12. Mohanty, R.K.; Thennarasu, S.; Mandal, A.B. Resveratrol stabilized gold nanoparticles enable surface loading of doxorubicin and anticancer activity. *Colloids Surf. B* **2014**, *114*, 138–143. [CrossRef]
13. Li, Y.; Dånmark, S.; Edlund, U.; Finne-Wistrand, A.; He, X.; Norgård, M.; Blomén, E.; Hultenby, K.; Andersson, G.; Lindgren, U. Resveratrol-conjugated poly-e-caprolactone facilitates in vitro mineralization and in vivo bone regeneration. *Acta Biomater.* **2011**, *7*, 751–758. [CrossRef]
14. Neo, Y.P.; Swift, S.; Ray, S.; Gizdavic-Nikolaidis, M.; Jin, J.; Perera, C.O. Evaluation of gallic acid loaded zein sub-micron electrospun fibre mats as novel active packaging materials. *Food Chem.* **2013**, *141*, 3192–3200. [CrossRef] [PubMed]
15. Ramírez-Ambrosi, M.; Caldera, F.; Trotta, F.; Berrueta, L.; Gallo, B. Encapsulation of apple polyphenols in β-CD nanosponges. *J. Incl. Phenom. Macrocycl. Chem.* **2014**, *80*, 85–92. [CrossRef]
16. Berlier, G.; Gastaldi, L.; Sapino, S.; Miletto, I.; Bottinelli, E.; Chirio, D.; Ugazio, E. MCM-41 as a useful vector for rutin topical formulations: Synthesis, characterization and testing. *Int. J. Pharm.* **2013**, *457*, 177–186. [CrossRef] [PubMed]
17. Chen, Y.; Lee, Y.D.; Vedala, H.; Allen, B.L.; Star, A. Exploring the chemical sensitivity of a carbon nanotube/green tea composite. *Acs Nano* **2010**, *4*, 6854–6862. [CrossRef]
18. Sousa, F.; Guebitz, G.M.; Kokol, V. Antimicrobial and antioxidant properties of chitosan enzymatically functionalized with flavonoids. *Process Biochem.* **2009**, *44*, 749–756. [CrossRef]
19. Božič, M.; Gorgieva, S.; Kokol, V. Homogeneous and heterogeneous methods for laccase-mediated functionalization of chitosan by tannic acid and quercetin. *Carbohydr. Polym.* **2012**, *89*, 854–864. [CrossRef] [PubMed]
20. Nunesa, C.; Maricato, É.; Cunha, Â.; Nunes, A.; da Silva, J.A.L.; Coimbra, M.A. Chitosan–caffeic acid–genipin films presenting enhanced antioxidant activity and stability in acidic media. *Carbohydr. Polym.* **2013**, *91*, 236–243. [CrossRef] [PubMed]

21. Trifković, K.T.; Milašinović, N.Z.; Djordjević, V.B.; Kruši, M.T.K.; Knežević-Jugović, Z.D.; Nedović, V.A.; Bugarski, B.M. Chitosan microbeads for encapsulation of thyme (*Thymus serpyllum* L.) polyphenols. *Carbohydr. Polym.* **2014**, *111*, 901–907. [CrossRef]
22. Moradi, M.; Tajik, H.; Rohani, S.M.R.; Oromiehie, A.R.; Malekinejad, H.; Aliakbarlu, J.; Hadian, M. Characterization of antioxidant chitosan film incorporated with Zataria multiflora Boiss essential oil and grape seed extract. *LWT-Food Sci. Technol.* **2012**, *46*, 477–484. [CrossRef]
23. Belščak-Cvitanović, A.; Stojanović, R.; Manojlović, V.; Komes, D.; Cindrić, I.J.; Nedović, V.; Bugarski, B. Encapsulation of polyphenolic antioxidants from medicinal plant extracts in alginate–chitosan system enhanced with ascorbic acid by electrostatic extrusion. *Food Res. Int.* **2011**, *44*, 1094–1101. [CrossRef]
24. Nagarajan, S.; Rami Reddy, B.S.; Tsibouklis, J. In vitro effect on cancer cells: Synthesis and preparation of polyurethane membranes for controlled delivery of curcumin. *J. Biomed. Mater. Res. Part A* **2011**, *99*, 410–417. [CrossRef]
25. Wu, H.; Wu, C.; He, Q.; Liao, X.; Shi, B. Collagen fiber with surface-grafted polyphenol as a novel support for Pd(0) nanoparticles: Synthesis, characterization and catalytic application. *Mater. Sci. Eng. C* **2010**, *30*, 770–776. [CrossRef]
26. Seshadri, S.; Sastry, T.P.; Jeevitha, D.; Samiksha, N. Synthesis and characterization of a novel bone graft material containing biphasic calcium phosphate and chitosan fortified with aloe vera. *Int. J. Drug Regul. Aff.* **2014**, *2*, 85–90.
27. Lin, F.H.; Dong, G.C.; Chen, K.S.; Jiang, G.J.; Huang, C.W.; Sun, J.S. Immobilization of Chinese herbal medicine onto the surface-modified calcium hydrogenphosphate. *Biomaterials* **2003**, *24*, 2413–2422. [CrossRef]
28. Sileika, T.S.; Barrett, D.G.; Zhang, R.; Lau, K.H.A.; Messersmith, P.B. Colorless multifunctional coatings inspired by polyphenols found in tea, chocolate, and wine. *Angew. Chem. Int. Ed.* **2013**, *52*, 10766–10770. [CrossRef] [PubMed]
29. Ferraris, S.; Zhang, X.; Prenesti, E.; Corazzari, I.; Turci, F.; Tomatis, M.; Vernè, E. Gallic acid grafting to a ferrimagnetic bioactive glass-ceramic. *J. Non-Cryst. Solids* **2016**, *432*, 167–175. [CrossRef]
30. Cazzola, M.; Corazzari, I.; Prenesti, E.; Bertone, E.; Vernè, E.; Ferraris, S. Bioactive glass coupling with natural polyphenols: Surface modification, bioactivity and anti-oxidant ability. *Appl. Surf. Sci.* **2016**, *367*, 237–248. [CrossRef]
31. Ferraris, S.; Spriano, S.; Bianchi, C.L.; Cassinelli, C.; Vernè, E. Surface modification of Ti-6Al-4 V alloy for biomineralization and specific biological response: Part II, alkaline phosphatase grafting. *J. Mater. Sci. Mater. Med.* **2011**, *22*, 1835–1842. [CrossRef] [PubMed]
32. Morra, M. Biochemical modification of titanium surfaces: Peptides and ECM proteins. *Eur. Cell. Mater.* **2006**, *12*, 1–15. [CrossRef]
33. Ferraris, S.; Spriano, S. Antibacterial titanium surfaces for medical implants. *Mater. Sci. Eng. C* **2016**, *61*, 965–978. [CrossRef] [PubMed]
34. Mohan, L.; Anandan, C.; Rajendran, N. Drug release characteristics of quercetin-loaded TiO_2 nanotubes coated with chitosan. *Int. J. Boil. Macromol.* **2016**, *93*, 1633–1638. [CrossRef] [PubMed]
35. Córdoba, A.; Satué, M.; Gómez-Florit, M.; Hierro-Oliva, M.; Petzold, C.; Lyngstadaas, S.P.; Gonzales-Martin, M.L.; Monjo, M.; Ramis, J.M. Flavonoid modified surfaces: Multifunctional bioactive biomaterials with osteopromotive, anti-inflammatory and anti-fibrotic potential. *Adv. Heathc. Mater.* **2015**, *4*, 540–549. [CrossRef] [PubMed]
36. Gurzawska, K.; Svava, R.; Yihua, Y.; Haugshøj, K.B.; Dirscherl, K.; Levery, S.B.; Byg, I.; Damager, I.; Nielsen, M.W.; Jørgensen, B.; Jørgensen, N.R.; Gotfredsen, K. Osteoblastic response to pectin nanocoating on titanium surfaces. *Mater. Sci. Eng. C* **2014**, *43*, 117–125. [CrossRef]
37. Erakovic, S.; Jankovic, A.; Tsui, G.; Tang, C.Y.; Miskovic-Stankovic, V.; Stevanovic, T. Novel bioactive antimicrobial lignin containing coatings on titanium obtained by electrophoretic deposition. *Int. J. Mol. Sci.* **2014**, *15*, 12294–12322. [CrossRef]
38. Džunuzović, E.S.; Džunuzović, J.V.; Marinković, A.D.; Marinović-Cincović, M.T.; Jeremić, K.B.; Nedeljković, J.M. Influence of surface modified TiO_2 nanoparticles by gallates on the properties of PMMA/TiO_2 nanocomposites. *Eur. Polym. J.* **2012**, *48*, 1385–1393. [CrossRef]
39. Verma, S.; Singh, A.; Mishra, A. Gallic acid: Molecular rival of cancer. *Environ. Toxicol. Pharmacol.* **2003**, *35*, 473–485. [CrossRef]
40. Lu, Z.; Nie, G.; Belton, P.S.; Tang, H.; Zhao, B. Structure–activity relationship analysis of antioxidant ability and neuroprotective effect of gallic acid derivatives. *Neurochem. Int.* **2006**, *48*, 263–274. [CrossRef]

41. *ASTM B348 Standard Specification for Titanium and Titanium Alloy Bars and Billets*; ASTM: West Conshohocken, PA, USA, 2010.
42. Spriano, S.; Vernè, E.; Ferraris, S. Multifunctional Titanium Surfaces for Bone Integration. EP Patent 2,214,732, 11 August 2010.
43. Ferraris, S.; Spriano, S.; Pan, G.; Venturello, A.; Bianchi, C.L.; Chiesa, R.; Faga, M.G.; Maina, G.; Vernè, E. Surface modification of Ti–6Al–4V alloy for biomineralization and specific biological response: Part I, inorganic modification. *J. Mater. Sci. Mater. Med.* **2011**, *22*, 533–545. [CrossRef]
44. Aita, H.; Hori, N.; Takeuchi, M.; Suzuki, T.; Yamada, M.; Anpo, M.; Ogawa, T. The effect of ultraviolet functionalization of titanium on integration with bone. *Biomaterials* **2009**, *30*, 1015–1025. [CrossRef]
45. Kokubo, T. Bioactive glass ceramics: Properties and applications. *Biomaterials* **1991**, *12*, 155–163. [CrossRef]
46. Singleton, V.L.; Rossi, J.A. Colorimetry of total phenolics with phosphomolybdic–phosphotungstic acid reagents. *Am. J. Enol. Vitic.* **1965**, *16*, 144–158.
47. Kokubo, T.; Takadama, H. How useful is SBF in predicting in vivo bone bioactivity? *Biomaterials* **2006**, *27*, 2907–2915. [CrossRef] [PubMed]
48. Friedman, M.; Jürgens, H.S. Effect of pH on the stability of plant phenolic compounds. *J. Agric. Food Chem.* **2000**, *48*, 2101–2110. [CrossRef] [PubMed]
49. Ejima, H.; Richardson, J.J.; Liang, K.; Best, J.P.; van Koeverden, M.P.; Such, G.K.; Caruso, F. One-step assembly of coordination complexes for versatile film and particle engineering. *Science* **2013**, *341*, 154–157. [CrossRef]
50. Surleva, A.; Atanasova, P.; Kolusheva, T.; Costadinnova, L. Study of the complex equilibrium between titanium (IV) and tannic acid. *J. Chem. Technol. Metall.* **2014**, *49*, 594–600.
51. Huguenin, J.; Hamady, S.O.; Bourson, P. Monitoring deprotonation of gallic acid by Raman spectroscopy. *J. Raman Spectrosc.* **2015**, *46*, 1062–1066. [CrossRef]
52. Mera, A.C.; Contreras, D.; Escalona, N.; Mansilla, H.D. BiOI microspheres for photocatalytic degradation of gallic acid. *J. Photochem. Photobiol. A* **2016**, *318*, 71–76. [CrossRef]
53. Araujo, P.Z.; Morando, P.J.; Blesa, M.A. Interaction of catechol and gallic acid with titanium dioxide in aqueous suspensions 1. Equilibrium studies. *Langmuir* **2005**, *21*, 3470–3474. [CrossRef]
54. Ball, V.; Meyer, F. Deposition kinetics and electrochemical properties of tannic acid on gold and silica. *Colloids Surf. A* **2016**, *491*, 12–17. [CrossRef]
55. Morra, M.; Cassinelli, C.; Buzzone, G.; Carpi, A.; DiSanti, G.; Giardino, R.; Fini, M. Surface chemistry effects of topographic modification of titanium dental implant surfaces: 1. Surface analysis. *Int. J. Oral Maxillofac. Implant.* **2003**, *18*, 40–45.
56. Prajatelistia, E.; Ju, S.W.; Sanandiya, N.D.; Jun, S.H.; Ahn, J.S.; Hwang, D.S. Tunicate-inspired gallic acid/metal ion complex for instant and efficient treatment of dentin hypersensitivity. *Adv. Healthc. Mater.* **2016**, *5*, 919–927. [CrossRef]
57. Yang, Z.; Wu, J.; Wang, X.; Wang, J.; Huang, N. Inspired chemistry for a simple but highly effective immobilization of vascular endothelial growth factor on GA functionalized plasma polymerized film Plasma Process. *Plasma Process. Polym.* **2012**, *9*, 718–725. [CrossRef]
58. Qiao, G.; Su, J.; He, M. Effect of (−)-epigallocatechin gallate on electrochemical behavior and surface film composition of Co–Cr alloy used in dental restorations. *Dent. Mater. J.* **2012**, *31*, 564–574. [CrossRef]
59. Kelemen, S.R.; Rose, K.D.; Kwiatek, P.J. Carbon aromaticity based on XPS II to II* signal intensity. *Appl. Surf. Sci.* **1993**, *64*, 167–174. [CrossRef]
60. Öteyaka, M.Ö.; Chevallier, P.; Robitaillec, L.; Larocheb, G. Effect of surface modification by ammonia plasma on vascular graft: PET film and PET scaffold. *Acta Phys. Pol. A* **2012**, *121*, 125–127. [CrossRef]
61. Ferraris, S.; Venturello, A.; Miola, M.; Cochis, A.; Rimondini, L.; Spriano, S. Antibacterial and bioactive nanostructured titanium surfaces for bone integration. *Appl. Surf. Sci.* **2014**, *311*, 279–291. [CrossRef]
62. Ferraris, S.; Bobbio, A.; Miola, M.; Spriano, S. Micro- and nano-textured, hydrophilic and bioactive titanium dental implants. *Surf. Coat. Technol.* **2015**, *276*, 374–383. [CrossRef]
63. Wang, H.; Lee, J.K.; Moursi, A.; Lannutti, J.J. Ca/P ratio effects on the degradation of hydroxyapatite in vitro. *J. Biomed. Mater. Res. Part A* **2003**, *67*, 599–608. [CrossRef]
64. Kulkarni, M.; Patil-Sen, Y.; Junkar, I.; Kulkarni, C.V.; Lorenzetti, M.; Iglič, A. Wettability studies of topologically distinct titanium surfaces. *Colloids Surf. B* **2015**, *129*, 47–53. [CrossRef]
65. Luxbacher, T. *The ZETA Guide: Principles of the Streaming Potential Technique*; Anton Paar GmbH: Graz, Austria, 2014.

66. Oc'wieja, M.; Adamczyk, Z.; Morga, M. Adsorption of tannic acid on polyelectrolyte monolayers determined in situ by streaming potential measurements. *J. Colloid Interface Sci.* **2015**, *438*, 249–258. [CrossRef] [PubMed]
67. Romdhane, A.; Aurousseau, M.; Guillet, A.; Mauret, E. Effect of pH and ionic strength on the electrical charge and particle size distribution of starch nanocrystal suspensions. *Starch-Stärke* **2015**, *67*, 319–327. [CrossRef]
68. Botelho, C.M.; LopesI, M.A.; Gibson, R.; Best, S.M.; Santos, J.D. Structural analysis of Si-substituted hydroxyapatite: Zeta potential and X-ray photoelectron spectroscopy. *J. Mater. Sci. Mater. Med.* **2002**, *13*, 1123–1127. [CrossRef] [PubMed]

© 2019 by the authors. Licensee MDPI, Basel, Switzerland. This article is an open access article distributed under the terms and conditions of the Creative Commons Attribution (CC BY) license (http://creativecommons.org/licenses/by/4.0/).

Article

Hydrothermal Synthesis of Protective Coating on Mg Alloy for Degradable Implant Applications

Jinshu Xie [1], Jinghuai Zhang [1,*], Shujuan Liu [2,*], Zehua Li [1], Li Zhang [1], Ruizhi Wu [1], Legan Hou [1] and Milin Zhang [1]

[1] Key Laboratory of Superlight Material and Surface Technology, Ministry of Education, College of Material Science and Chemical Engineering, Harbin Engineering University, Harbin 150001, China; 2013105235@hrbeu.edu.cn (J.X.); 1352091844@hrbeu.edu.cn (Z.L.); 2013xgd@hrbeu.edu.cn (L.Z.); rzwu@hrbeu.edu.cn (R.W.); houlegan@hrbeu.edu.cn (L.H.); zhangmilin@hrbeu.edu.cn (M.Z.)
[2] Department of Materials Physics and Chemistry, Harbin Institute of Technology, Harbin 150001, China
* Correspondence: zhangjinghuai@hrbeu.edu.cn (J.Z.); liusj0817@hit.edu.cn (S.L.); Tel.: +86-451-8256-9295 (J.Z.)

Received: 26 January 2019; Accepted: 26 February 2019; Published: 28 February 2019

Abstract: Biodegradable magnesium (Mg) alloys are known as "the new generation of biomedical metal materials". However, high degradation rates restrict their clinical application. To overcome this issue, a new and simple method for producing of protective coating based on hydrothermal synthesis at 200 °C in 0.5 M $NaHCO_3$ was elaborated. The microstructure, elemental and phase composition of the produced films were examined by scanning electron microscope (SEM), X-ray energy-dispersive spectrometer (EDS) and X-ray diffraction (XRD). The mechanical strength of the protective coating was evaluated by grid scribing method. The corrosion protection effect was evaluated using linear sweep voltammogram (LSV) and electrochemical impedance spectroscopy (EIS) methods in the simulated body fluid (SBF). Since the corrosion process is accompanied by stoichiometric evolution of hydrogen, the amount of the latter was measured to quantify the overall corrosion rate. Both the coatings morphology and phase composition were sensitive to the treatment duration. The coating formed after 0.5 h was loose and mainly consisted of spherical flower-like $Mg_5(CO_3)_4(OH)_2·4H_2O$ accompanied by small amounts of $Mg(OH)_2$. The treatment duration of 3 h resulted in a thicker compact coating composed mainly of irregular granular $MgCO_3$ as well as $Mg(OH)_2$. The coating providing the most effective protection and uniform corrosion was achieved by 2 h treatment at 200 °C.

Keywords: Mg alloy; corrosion protection; hydrothermal synthesis; coating; degradable implant

---

1. Introduction

Mg alloys have attracted much attention for their potential use as a new class of biodegradable medical implant materials, as they possess good biocompatibility [1]. Mg alloy are biodegradable, and their degradation products can be excreted through human metabolism. Furthermore, Mg is an essential element for human organisms, which could have stimulatory effects on the bone grafting. The density and the elastic modulus of Mg alloys are close to those of natural bones [2]. Therefore, Mg alloys are known as "the new generation of biomedical metal materials" [3]. The biological applications of Mg alloys are limited mainly due to their too rapid degradation rate for most implanted devices [4,5]. Although some studies related to the corrosion/degradation behavior of Mg and Mg alloys have been intensively carried out [6], further research is still needed to improve their corrosion characteristics and surface properties.

To improve the corrosion resistance of Mg alloys, surface modification has become the research focus of medical Mg alloys [7]. Chemical conversion coatings [8,9], electrochemical coatings [10,11], polymer coatings [12,13], ceramic coatings [14,15], ion implantation coatings [16,17], composite coatings [18,19] etc. improve the corrosion performance of Mg alloys. The drawbacks of these methods are often complicated process, limited corrosion protection, poor biocompatibility and poor adhesion [20–24]. While passivation of the Mg surface by oxidation is a very simple process, the spontaneously formed oxide coating is loose and does not provide effective corrosion protection. Alkaline heat treatment, however, is a simple method which effectively improves the corrosion resistance of Mg alloys because of the compact oxide film that forms [25]. As a special chemical method, hydrothermal treatment has been gradually introduced into the corrosion prevention of Mg alloys in recent years [26,27]. In this study, we applied a hydrothermal treatment in aqueous $NaHCO_3$ solution to the high-performance biomedical $Mg_6Ho_{0.5}Zn$ alloy.

The selected substrate $Mg_{6.659}Ho_{0.536}Zn_{0.005}Fe_{0.002}Ni_{0.007}Cu$, wt.% extruded alloy, is a promising biodegradable medical material [28]. Nano-spaced and solute-segregated basal plane stacking faults (SFs) with an extremely large aspect ratio, can be formed and distributed uniformly throughout the fine dynamic recrystallized (DRXed) grains in the as-extruded $Mg_6Ho_{0.5}Zn$ alloy by adding Ho and Zn and controlling their ratio as well as appropriate preparation processing parameters. Compared with alloys possessing classic microstructure, the new alloy with profuse SFs exhibits uniform in-vitro and in-vivo corrosion behavior, low corrosion rate and optimal mechanical properties. In addition, the $Mg_6Ho_{0.5}Zn$ alloy shows no significant toxicity to Saos-2 cells.

In this paper, the effect of the hydrothermal treatment duration on the morphology, composition, mechanical and electrochemical properties of the protective coating formed on the $Mg_6Ho_{0.5}Zn$ substrate surface were investigated. It should be noted that the hydrothermal treatment may not be limited only to the $Mg_6Ho_{0.5}Zn$ alloy, but might be suitable for other Mg alloys as well.

2. Materials and Methods

2.1. Samples Preparation

The alloy with nominal composition $Mg_{6.659}Ho_{0.536}Zn_{0.005}Fe_{0.002}Ni_{0.007}Cu$, wt.% and its special structure and outstanding corrosion behavior was reported [29]. The substrate samples were ground and polished with SiC abrasive papers from P120 down to P2000 to achieve a smooth surface and get rid of impurity layer and then degreased with ethanol and acetone (1:1) in ultrasonic bath for 10 min, followed by drying in the air.

Four sets of samples were prepared by hydrothermal treatment in 0.5 M $NaHCO_3$ aqueous solution at 200 °C in the autoclave for 0.5, 1, 2 and 3 h, followed by drying at 200 °C for 1 h. These hydrothermally treated samples were designated as HT0.5, HT1, HT2 and HT3 respectively.

2.2. Characterization

The morphology and phase composition of the coatings were examined by scanning electron microscope (SEM, Quanta 200 FEG, FEI, Hillsboro, OR, USA) equipped with X-ray energy-dispersive spectrometer (EDS, Quanta 200 FEG, FEI, Hillsboro, OR, USA) and X-ray diffraction (XRD, D/MAX-Ultima III, Rigaku Corporation, Tokyo, Japan).

To test the mechanical strength of the coating, a grid was scribed on the sample surface using a sharp blade and controlled by SEM.

Linear sweep voltammogram (LSV) and electrochemical impedance spectroscopy (EIS) measurements were conducted in the simulated body fluid (SBF) using PGSTAT302N potentiostat (Metrohm Autolab, Utrecht, The Netherlands). The composition of the SBF is depicted in Table 1. The SBF was buffered with Tris-HCl and the pH 7.4 was set. A three-electrode cell set-up was used, with the tested sample as the working electrode (WE), a platinum wire as the counter electrode (CE) and a saturated calomel electrode (SCE) as the reference electrode (RE). Prior to electrochemical measurements, samples were kept in the SBF for 0.5 h. LSV was conducted in the potential range from −300 to 300 mV vs open circuit potential (OCP) at 2 mV/s. EIS was performed at OCP, and 10 mV_{RMS} AC excitation signal amplitude in the frequency range from 100 kHz to 10 mHz.

Table 1. Composition of the simulated body fluid (SBF).

Component	Mass	Degree of Purity [%]
NaCl	8.035 g/L	99.5
NaHCO$_3$	0.355 g/L	99.5
KCl	0.225 g/L	99.5
K$_2$HPO$_4$·3H$_2$O	0.231 g/L	99.0
MgCl$_2$·6H$_2$O	0.311 g/L	98.0
C(HCl) = 1 M	39 mL	–
CaCl$_2$	0.292 g/L	95.0
Na$_2$SO$_4$	0.072 g/L	99.0
Tris	6.118 g/L	99.0

Long-term corrosion behavior was evaluated by immersion in SBF at 37 ± 0.5 °C for 5 days. The evolved hydrogen was continuously collected by means of the funnel and the burette above the sample. To compensate the pH shift due to hydrogen evolution, the SBF was replaced every 36 h and the SBF amount was controlled to be 30 mL/cm^2 of the sample surface.

3. Results and Discussion

3.1. Morphology and Composition

Figure 1 shows SEM micrographs of the four samples after hydrothermal treatment with different durations. Two distinguishable morphologies were formed: the coating of the samples HT0.5 and HT1 consisted of numerous spheres with the diameter ranging from 4 to 10 μm. Smaller spheres composed compactly aggregated layer, while larger spheres were scattered on the compact layer surface (Figure 1a–d). The spheres itself are composed of thin nano-flakes. Remarkably, the nano-flakes grew together forming rather dense spherical particles once the treatment duration was prolonged. The morphology was changed cardinally after 2 h of the treatment (sample HT2—Figure 1e,f): the surface was composed of numerous irregular particles of about 3–5 μm. The compactness and uniformity of the coating were also enhanced. Further prolongation of the synthesis time does not change the morphology very much: the film got denser and more compact (sample HT3—Figure 1g,h).

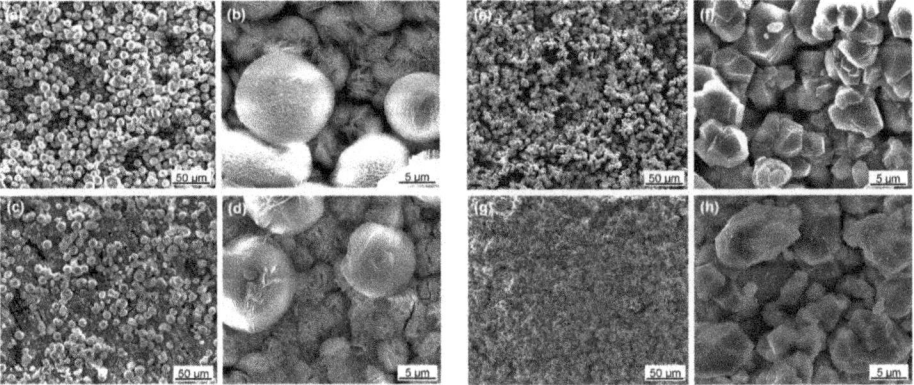

Figure 1. SEM micrographs of samples HT0.5 (**a,b**), HT1 (**c,d**), HT2 (**e,f**) and HT3 (**g,h**) prepared by hydrothermal treatment in 0.5 M NaHCO$_3$ at 200 °C for 0.5, 1, 2 and 3 h respectively.

HT0.5 and HT1 mainly consisted of Mg$_5$(CO$_3$)$_4$(OH)$_2$·4H$_2$O, as it can be seen from XRD patterns (Figure 2). Diffraction peaks of these two samples also indicated the presence of small Mg(OH)$_2$ quantities. In contrast to this, the main phase composition of the HT2 and HT3 samples was MgCO$_3$

with a few Mg(OH)$_2$ inclusions. Thus, the duration of the hydrothermal treatment influences both the morphology and the composition of the formed coating.

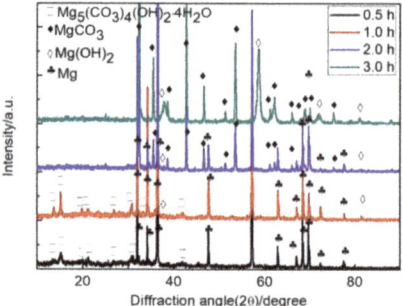

Figure 2. XRD patterns of samples HT0.5, HT1, HT2 and HT3 prepared by hydrothermal treatment in 0.5 M NaHCO$_3$ at 200 °C for 0.5, 1, 2 and 3 h respectively.

The film thickness build-up can also be followed by XRD patterns. Thus, the intensity of Mg diffraction peaks, which come from the substrate alloy, are roughly the same for the HT0.5 and HT1 samples, indicating, that the coating thickness is also comparable. In the case of the HT2 sample, the intensity of the Mg diffraction peaks drops significantly, indicating increase of the coating thickness, as well as its compactness. It is worth noting that the different phase may have different X-ray absorption behavior. Finally, the HT3 sample shows no Mg diffraction peaks, thus the formed coating is thicker, than that of the HT2 sample. These observations also agree with the morphology change detected by SEM.

To support the XRD data, elemental analysis of the HT2 sample was carried out by SEM-EDS, as shown in the Figure 3a,b. The C/O atom ratio was 1:2.25, which roughly represents the MgCO$_3$ phase detected by XRD. The cross-sectional morphology of this hydrothermal film was also observed by SEM, as shown in the Figure 3c and the coating thickness amounted to 15–20 µm. Considering an in-situ formation of the coating, a good adhesion of the carbonate layer to the substrate alloy can be assumed. This is indirectly confirmed by the absence of cracks and caves.

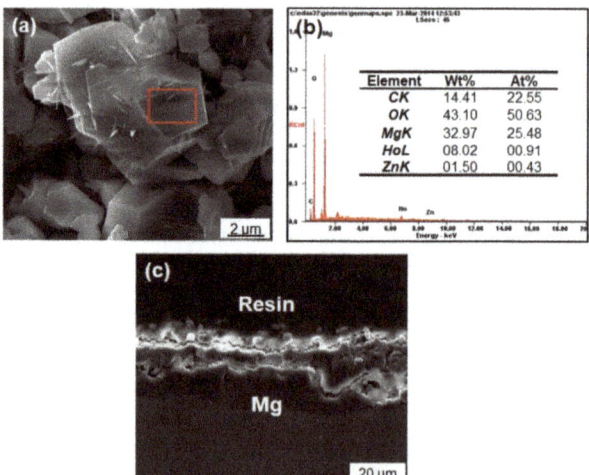

Figure 3. (a) SEM micrograph; (b) elemental composition; and (c) cross–section micrograph of the HT2 sample.

3.2. Formation Mechanism

Based on the observations made, the formation of the coating can be divided into three stages: nucleation, transformation and densification, as schematically shown in the Figure 4. First, Mg^{2+} and OH^- are formed as products of redox reaction between water and magnesium metal with corresponding half-reactions:

$$2H_2O + 2e^- \rightarrow H_2 + 2OH^-$$

$$Mg - 2e^- \rightarrow Mg^{2+}$$

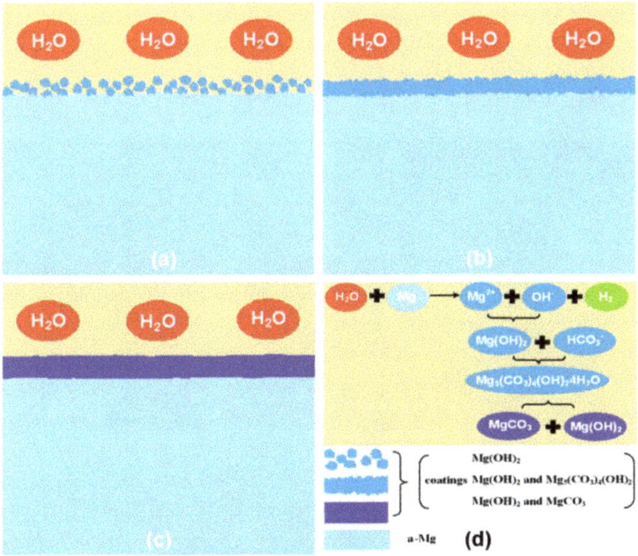

Figure 4. Schematic representation of the suggested coating formation mechanism: (**a**) nucleation; (**b**) transformation; (**c**) densification; (**d**) overall coating formation mechanism.

The rates of these reactions are crucial in defining the overall coating nucleation rate:

$$Mg^{2+} + 2OH^- \rightarrow Mg(OH)_2$$

Since the reaction conditions accelerate the Mg oxidation by water, the nucleation rate is rather high, which results in the loose spherical $Mg(OH)_2$ particles (Figure 4a). Since the formed coating possesses many channels, the electrolyte can penetrate it and stands in the direct contact to the metallic substrate. As a result, such coating cannot provide enough protection against corrosion. During the transformation phase (Figure 4b) the carbonate ions partially substitute hydroxide forming mixed hydroxycarbonate:

$$5Mg(OH)_2 + 4HCO_3^- \rightarrow Mg_5(CO_3)_4(OH)_2 \cdot 4H_2O + 4OH^-$$

The transformation of the coating may occur in the whole bulk of the $Mg(OH)_2$ as the non-converted coating can be penetrated by the electrolyte as mentioned above. The hydroxycarbonate carbonate decomposes due to its instability:

$$Mg_5(CO_3)_4(OH)_2 \cdot 4H_2O \rightarrow 4MgCO_3 + Mg(OH)_2 + 4H_2O$$

forming magnesium carbonate and magnesium hydroxide, while the latter, again, can react with carbonate ions. The $MgCO_3$ layer is compact and uniform.

3.3. Mechanical Strength

The coating compactness increases with the treatment duration, however, there is a trend off between the compactness, which is crucial for the efficient corrosion protection, and mechanical strength, which is important for the long-term stability of the coating. The adhesion of the coating was tested for the samples HT2 and HT3 according to the method described in the experimental section. Figure 5 depicts SEM micrographs of the HT2 (Figure 5a,b) and HT3 (Figure 5c,d) samples after the strength test. The HT2 coating remained unchanged, maintaining its compact structure. No further damages of the coating could be found at the cut sites, thus it can be concluded that the coating exhibits sufficient mechanical strength and adhesion to the metal substrate. This is not the case for the HT3 sample, where obvious delamination occurred. Thus, it is clear that the thicker coating loses its mechanical strength, followingly losing its corrosion protective properties. Based on this, the hydrothermal treatment duration of 2 h is the optimum treatment duration.

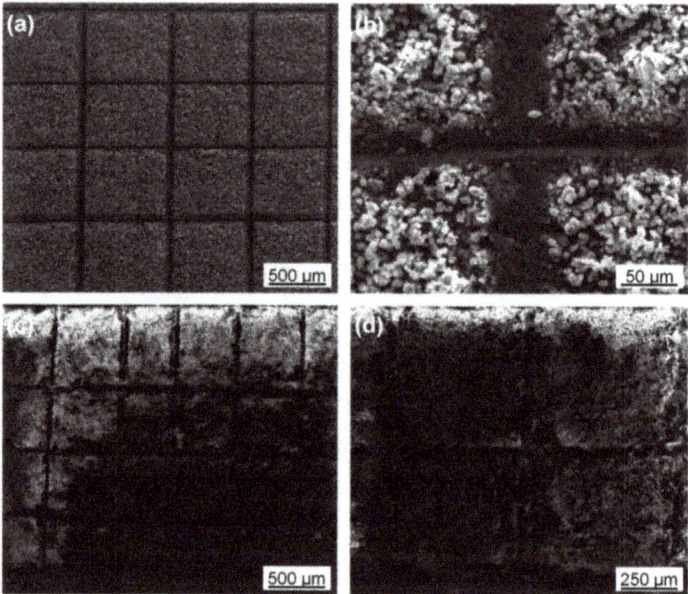

Figure 5. SEM micrographs of the samples HT2 (**a,b**) and HT3 (**c,d**) after adhesion test.

3.4. Electrochemistry

Break-through potential of the samples was determined by means of LSV. Prior to polarization, samples were conditioned at OCP for 0.5 h. The break-through potential (E_{corr}) and corrosion current density (J_{corr}) were derived from the Tafel-type plot (Figure 6a). The derived data is summarized in Table 2.

Table 2. J_{corr} and E_{corr} of the samples HT0.5, HT1, HT2 and HT3.

Hydrothermal Time [h]	J_{corr} [A/cm^2]	E_{corr} [V]
0	2.015×10^{-4}	−1.793
0.5	1.114×10^{-4}	−1.854
1	6.765×10^{-5}	−1.712
2	2.129×10^{-5}	−1.450
3	2.896×10^{-5}	−1.367

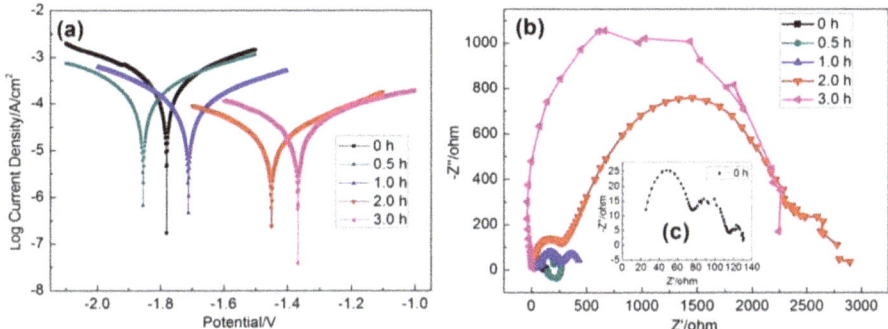

Figure 6. (a) Polarization curves and (b,c) Nyquist plots of untreated samples and samples HT0.5, HT1, HT2 and HT3 prepared by hydrothermal treatment in 0.5 M NaHCO$_3$ at 200 °C for 0.5, 1, 2 and 3 h respectively.

The breakthrough potential shifts in the anodic direction with increased treatment time except for the HT0.5. The same trend is valid for the corrosion current density. The corrosion current is caused by Mg oxidation and is, thus, a good indicator of corrosion stability: as less the current density is, as more efficient the correction protection is. Therefore, the corrosion rate of the HT3 sample was 14 times less, than that of the untreated substrate and the breakthrough potential shifted by 0.426 V anodically.

To support the LSV data, charge transfer resistance (R_{ct}) values were evaluated by means of EIS. Figure 6b,c shows Nyquist plots of the untreated substrate and samples HT0.5, HT1, HT2 and HT3. Samples HT1 and HT2 have at least two time constants: the high frequency capacitive loop is associated with R_{ct} and the low frequency loop is associated with electrolyte penetration, including uptake of water and intrusion of ions. Spectrum of the HT0.5 sample additionally includes pseudo-inductive loop at low frequencies. It can be related with adsorption of Mg$^+$ or Mg^{2+} at the Mg surface, thus indicating corroded sample surface. Without comprehensive analysis of the EIS data it can be indicated, that the R_{ct} greatly increases with increased sample treatment duration, which in turn, is a sign of enhanced corrosion protection [30–32]. The sample HT3 does only show one time constant, which suggests that the coating is compact enough, so that no electrolyte penetration is detectable on this way.

3.5. Immersion Corrosion of Coating

Since the sample HT2 showed satisfactory short-term performance and stability, it was chosen for the long-term performance test in SBF at 37 °C for 5 days. The experiment was conducted as described in the experimental section. The obtained data is shown in the Figure 7. The overall corrosion reaction of Mg at its free corrosion potential can be expressed as follows [33,34]:

$$Mg + 2H_2O \rightarrow Mg^{2+} + 2OH^- + H_2$$

Thus, one mole of the oxidized Mg corresponds to one mole of evolved hydrogen and measuring its volume enables determining of the Mg weight loss. The quantity of the dissolved hydrogen was neglected.

After 120 h the quantity of the evolved hydrogen reached a mark at 0.7 mL/cm^2, whereas that of the HT2 was slightly above 0.2 mL/cm^2. This proves the fact that the coated sample also shows resistance against corrosion in the course of a long-term corrosive load. The corrosion rate of the untreated sample was constant during the experiment duration of 120 h, while that of the HT2 sample was continuously decreasing. The drop of the corrosion rate in the case of the HT2 may indicate additional surface passivation or local establishment of conditions, which slow down the kinetics of the Mg oxidation. Figure 8a,b shows the HT2 coating morphology after the long-term test. It remained almost unchanged as compared to the freshly prepared sample (Figure 1e,f): the coating is still uniform and compact completely covering the metal surface with no voids or cracks.

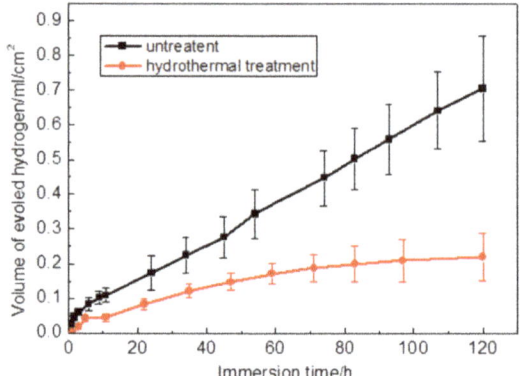

Figure 7. Amount of evolved hydrogen as function of time for HT2 and an untreated sample.

Figure 8. SEM micrographs of the samples HT2 (**a**,**b**) after static immersion in SBF solution at 37 °C for 5 days.

4. Conclusions

In the present study, we proposed a simple, fast and effective method of biomedical $Mg_6Ho_{0.5}Zn$ alloy protection by means of in-situ hydrothermal growth of protective coating in 0.5 M $NaHCO_3$ at 200 °C. The morphology and the phase composition of such grown coating may be influenced by varying the treatment duration. A thin film mainly consisting of spherical flower-like $Mg_5(CO_3)_4(OH)_2·4H_2O$ particles resulted after 0.5 h of the hydrothermal treatment. Increasing the treatment duration up to 3 h results in a compact thick film consisting of irregular granular $MgCO_3$ as well as some $Mg(OH)_2$. By means of comprehensive evaluation of the mechanical and electrochemical coating properties, the treatment duration of 2 h was found to be the optimum treatment duration. Longer processing leads to deterioration of the coating mechanical properties. The protective coating may slow down the penetration of detrimental anions, such as Cl^-, thus making the alloy corrode uniformly and slowly in-vivo as well.

Author Contributions: Conceptualization, J.Z. and S.L.; Methodology and Software J.X., Z.L. and L.Z.; Investigation and Data Curation, R.W., L.H. and M.Z.; Writing—Original Draft Preparation, J.X. and L.Z.; Writing—Review and Editing and Supervision, J.Z.

Funding: The research was funded by National Natural Science Foundation of China (No. 51871069), Natural Science Foundation of Heilongjiang Province of China (No. E2017030) and Foundation of State Key Laboratory of Rare Earth Resources Utilization (No. RERU2018017).

Conflicts of Interest: The authors declare no conflict of interest.

References

1. Zhao, D.; Witte, F.; Lu, F.; Wang, J.; Li, J.; Qin, L. Current status on clinical applications of magnesium-based orthopedic implants: A review from clinical translational perspective. *Biomaterials* **2017**, *112*, 287–302. [CrossRef] [PubMed]
2. Husak, Y.; Solodovnyk, O.; Yanovska, A.; Kozik, Y.; Liubchak, I.; Ivchenko, V.; Mishchenko, O.; Zinchenko, Y.; Kuznetsov, V.; Pogorielov, M. Degradation and in vivo response of hydroxyapatite-coated Mg alloy. *Coatings* **2018**, *8*, 375. [CrossRef]
3. Tian, P.; Liu, X. Surface modification of biodegradable magnesium and its alloys for biomedical applications. *Regen. Biomater.* **2015**, *2*, 135–151. [CrossRef] [PubMed]
4. Sanchez, A.H.M.; Luthringer, B.J.; Feyerabend, F.; Willumeit, R. Mg and Mg alloys: How comparable are in vitro and in vivo corrosion rates: A review. *Acta. Biomater.* **2015**, *13*, 16–31. [CrossRef] [PubMed]
5. Han, P.; Cheng, P.; Zhang, S.; Zhao, C.; Ni, J.; Zhang, Y.; Zhong, W.; Hou, P.; Zhang, X.; Zheng, Y.; et al. In vitro and in vivo studies on the degradation of high-purity Mg (99.99 wt.%) screw with femoral intracondylar fractured rabbit model. *Biomaterials* **2015**, *64*, 57–69. [CrossRef] [PubMed]
6. Esmaily, M.; Svensson, J.E.; Fajardo, S.; Birbilis, N.; Frankel, G.S.; Virtanen, S.; Arrabal, R.; Thomas, S.; Johansson, L.G. Fundamentals and advances in magnesium alloy corrosion. *Prog. Mater. Sci.* **2017**, *89*, 92–193. [CrossRef]
7. Gu, X.N.; Li, S.S.; Li, X.M.; Fan, Y.B. Magnesium based degradable biomaterials: A review. *Front. Mater. Sci.* **2014**, *8*, 200–318. [CrossRef]
8. Pan, C.J.; Pang, L.Q.; Hou, Y.; Lin, Y.B.; Gong, T.; Liu, T.; Ye, W.; Ding, H.Y. Improving corrosion resistance and biocompatibility of magnesium alloy by sodium hydroxide and hydrofluoric acid treatments. *Appl. Sci.* **2016**, *7*, 33. [CrossRef]
9. Sun, W.; Zhang, G.; Tan, L.; Yang, K.; Ai, H. The fluoride coated AZ31B magnesium alloy improves corrosion resistance and stimulates bone formation in rabbit model. *Mater. Sci. Eng. C* **2016**, *63*, 506–511. [CrossRef] [PubMed]
10. Zhou, Y.R.; Zhang, S.; Nie, L.L.; Zhu, Z.J.; Zhang, J.Q.; Cao, F.H.; Zhang, J.X. Electrodeposition and corrosion resistance of Ni–P–TiN composite coating on AZ91D magnesium alloy. *Trans. Nonferrous Met. Soc. China* **2016**, *26*, 2976–2987. [CrossRef]
11. Liu, Y.; Xue, J.; Luo, D.; Wang, H.; Gong, X.; Han, Z.; Ren, L. One-step fabrication of biomimetic superhydrophobic surface by electrodeposition on magnesium alloy and its corrosion inhibition. *J. Colloid. Interface. Sci.* **2017**, *491*, 313–320. [CrossRef] [PubMed]
12. Badruddoza Dithi, A.; Nezu, T.; Nagano-Takebe, F.; Hasan, M.; Saito, T.; Endo, K. Application of solution plasma surface modification technology to the formation of thin hydroxyapatite film on titanium implants. *Coatings* **2019**, *9*, 3. [CrossRef]
13. Liu, J.; Xi, T. Enhanced anti-corrosion ability and biocompatibility of PLGA coatings on Mg–Zn–Y–Nd alloy by BTSE-APTES pre-treatment for cardiovascular stent. *J. Mater. Sci. Technol.* **2016**, *32*, 845–857. [CrossRef]
14. Çelik, İ. Structure and surface properties of Al_2O_3-TiO_2 ceramic coated AZ31 magnesium alloy. *Ceram. Int.* **2016**, *42*, 13659–13663. [CrossRef]
15. Ji, M.K.; Oh, G.; Kim, J.W.; Park, S.; Yun, K.D.; Bae, J.C.; Lim, H.P. Effects on antibacterial activity and osteoblast viability of non-thermal atmospheric pressure plasma and heat treatments of TiO_2 nanotubes. *J. Nanosci. Nanotechnol.* **2017**, *17*, 2312–2315. [CrossRef] [PubMed]
16. Jin, W.; Wu, G.; Feng, H.; Wang, W.; Zhang, X.; Chu, P.K. Improvement of corrosion resistance and biocompatibility of rare-earth WE43 magnesium alloy by neodymium self-ion implantation. *Corros. Sci.* **2015**, *94*, 142–155. [CrossRef]
17. Cheng, M.; Qiao, Y.; Wang, Q.; Qin, H.; Zhang, X.; Liu, X. Dual ions implantation of zirconium and nitrogen into magnesium alloys for enhanced corrosion resistance, antimicrobial activity and biocompatibility. *Colloids. Surf. B Biointerfaces* **2016**, *148*, 200–210. [CrossRef] [PubMed]
18. Wei, Z.; Tian, P.; Liu, X.; Zhou, B. In-vitro, degradation, hemolysis, and cytocompatibility of PEO/PLLA composite coating on biodegradable AZ31 alloy. *J. Biomed. Mater. Res. B Appl. Biomater.* **2015**, *103*, 342–354. [CrossRef] [PubMed]
19. Tian, P.; Xu, D.; Liu, X. Mussel-inspired functionalization of PEO/PCL composite coating on a biodegradable AZ31 magnesium alloy. *Colloids. Surf. B Biointerfaces* **2016**, *141*, 327–337. [CrossRef] [PubMed]

20. Tian, P.; Liu, X.; Ding, C. In vitro degradation behavior and cytocompatibility of biodegradable AZ31 alloy with PEO/HT composite coating. *Colloids. Surf. B Biointerfaces* **2015**, *128*, 44–54. [CrossRef] [PubMed]
21. Dong, K.; Song, Y.; Shan, D.; Han, E.H. Corrosion behavior of a self-sealing pore micro-arc oxidation film on AM60 magnesium alloy. *Corros. Sci.* **2015**, *100*, 275–283. [CrossRef]
22. Wu, G.; Gong, L.; Feng, K.; Wu, S.; Zhao, Y.; Chu, P.K. Rapid degradation of biomedical magnesium induced by zinc ion implantation. *Mater. Lett.* **2011**, *65*, 661–663. [CrossRef]
23. Amaravathy, P.; Rose, C.; Sathiyanarayanan, S.; Rajendran, N. Evaluation of in vitro bioactivity and MG63 oesteoblast cell response for TiO coated magnesium alloys. *J. Sol-Gel. Sci. Technol.* **2012**, *64*, 694–703. [CrossRef]
24. Zhao, J.; Chen, L.J.; Yu, K.; Chen, C.; Dai, Y.L.; Qiao, X.Y.; Yan, Y.; Yu, Z.M. Effects of chitosan coating on biocompatibility of Mg–6%Zn–10%Ca$_3$(PO$_4$)$_2$ implant. *Trans. Nonferrous Met. Soc. China* **2015**, *25*, 824–831. [CrossRef]
25. Gu, X.N.; Zheng, W.; Cheng, Y.; Zheng, Y.F. A study on alkaline heat treated Mg–Ca alloy for the control of the biocorrosion rate. *Acta Biomater.* **2009**, *5*, 2790–2799. [CrossRef] [PubMed]
26. Zhu, Y.; Wu, G.; Zhang, Y.H.; Zhao, Q. Growth and characterization of Mg(OH)$_2$ film on magnesium alloy AZ31. *Appl. Surf. Sci.* **2011**, *257*, 6129–6137. [CrossRef]
27. Peng, F.; Li, H.; Wang, D.; Tian, P.; Tian, Y.; Yuan, G.; Xu, D.; Liu, X. Enhanced corrosion resistance and biocompatibility of magnesium alloy by Mg–Al-layered double hydroxide. *ACS. Appl. Mater. Interfaces.* **2016**, *8*, 35033–36044. [CrossRef] [PubMed]
28. Zhang, L.; Zhang, J.; Xu, C.; Jing, Y.; Zhuang, J.; Wu, R.; Zhang, M. Formation of stacking faults for improving the performance of biodegradable Mg–Ho–Zn alloy. *Mater. Lett.* **2014**, *133*, 158–162. [CrossRef]
29. Jiao, Y.; Zhang, J.; Kong, P.; Zhang, Z.; Jing, Y.; Zhuang, J.; Wang, W.; Zhang, L.; Xu, C.; Wu, R.; et al. Enhancing the performance of Mg-based implant materials by introducing basal plane stacking faults. *J. Mater. Chem. B* **2015**, *3*, 7386–7400. [CrossRef]
30. Dan, S.; Ma, A.B.; Jiang, J.; Lin, P.; Yang, D.; Fan, J. Corrosion behavior of equal-channel-angular-pressed pure magnesium in NaCl aqueous solution. *Corros. Sci.* **2010**, *52*, 481–490.
31. Yao, C.; Lv, H.; Zhu, T.; Zheng, W.; Yuan, X.; Gao, W. Effect of Mg content on microstructure and corrosion behavior of hot dipped Zn–Al–Mg coatings. *J. Alloys. Compd.* **2016**, *670*, 239–248. [CrossRef]
32. Mosiałek, M.; Mordarski, G.; Nowak, P.; Simka, W.; Nawrat, G.; Hanke, M.; Socha, R.P.; Michalska, J. Phosphate–permanganate conversion coatings on the AZ81 magnesium alloy: SEM, EIS and XPS studies. *Surf. Coat. Technol.* **2011**, *206*, 51–62. [CrossRef]
33. Zhou, W.; Shan, D.; Han, E.H.; Ke, W. Structure and formation mechanism of phosphate conversion coating on die-cast AZ91D magnesium alloy. *Corros. Sci.* **2008**, *50*, 329–337. [CrossRef]
34. Lu, F.; Ma, A.; Jiang, J.; Guo, Y.; Yang, D.; Song, D.; Chen, J. Significantly improved corrosion resistance of heat-treated Mg–Al–Gd alloy containing profuse needle-like precipitates within grains. *Corros. Sci.* **2015**, *94*, 171–178. [CrossRef]

© 2019 by the authors. Licensee MDPI, Basel, Switzerland. This article is an open access article distributed under the terms and conditions of the Creative Commons Attribution (CC BY) license (http://creativecommons.org/licenses/by/4.0/).

MDPI
St. Alban-Anlage 66
4052 Basel
Switzerland
Tel. +41 61 683 77 34
Fax +41 61 302 89 18
www.mdpi.com

Coatings Editorial Office
E-mail: coatings@mdpi.com
www.mdpi.com/journal/coatings

www.ingramcontent.com/pod-product-compliance
Lightning Source LLC
LaVergne TN
LVHW070601100526
838202LV00012B/533